# Riddle of the Ice

ANCHOR BOOKS

DOUBLEDAY

New York London Toronto Sydney Auckland

# RIDDLE OF THE
# ICE

**A SCIENTIFIC ADVENTURE
INTO THE ARCTIC**

# Myron Arms

AN ANCHOR BOOK
PUBLISHED BY DOUBLEDAY
a division of Bantam Doubleday Dell Publishing Group, Inc.
1540 Broadway, New York, New York 10036

ANCHOR BOOKS, DOUBLEDAY, and the portrayal of an anchor are trademarks
of Doubleday, a division of Bantam Doubleday Dell Publishing Group, Inc.

Book design by James Sinclair

*Library of Congress Cataloging-in-Publication Data*
Arms, Myron.
Riddle of the ice: a scientific adventure into the Arctic / by Myron Arms.
—1st Anchor Books ed.
p.     cm.
Includes bibliographical references.
1. Arms, Myron.   2. Brendan's Isle (Sailboat)   3. Arctic regions—Description
and travel.   4. Sea ice—Arctic regions.   5. Climate changes—Arctic regions.
I. Title.
G743.A698   1998
919.8—dc21   97-26104
CIP

ISBN 0-385-49092-5
Copyright © 1998 by Myron Arms

for Kay

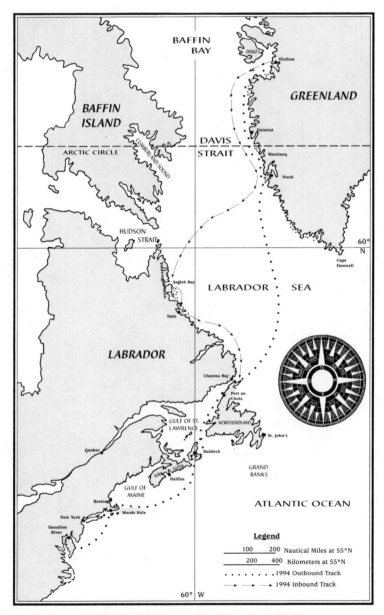

# Sailing Route: The Davis Strait

# Contents

"Perhaps I had even then begun to want to understand the ice . . . in an attempt to recapture something we have lost."

—Peter Hoeg,
*Smilla's Sense of Snow*

# A Note from the Author

In the summer of 1991 a crew of young sailors and I set out on a three-month voyage aboard a fifty-foot sailing cutter, *Brendan's Isle,* from the east coast of the United States to the remote Tourngat region of northern Labrador. As with any long sailing voyage, this one was full of surprises, and my crew and I returned home in September with both less and more than we had bargained for.

Less because the sailboat was stopped along the south Labrador coast by a massive area of sea ice that should not have been there and that forced her to turn around six hundred kilometers south of her intended destination. More because the ice led to a question, the question became a riddle, and the riddle evolved into a series of encounters with a group of scientists struggling to understand the Arctic ice regime and its relationship to changing global climate.

Haunted by the seemingly inexplicable mass of sea ice that had stopped *Brendan's Isle,* I returned home in the fall of 1991 determined to learn whatever I could about the geophysical forces at work in the ice. Why, during one of the warmest years on record, had the ice been there at all? Where had it come from? How was it related to other changes taking place in the Arctic and in the climate system as a whole?

Questions that at first seemed simple, however, soon grew more complex, each one opening, layer upon layer, into new and surprising areas of scientific inquiry. What started as a simple follow-up call to the Canadian Ice Centre in Ottawa soon grew into a series of contacts with climatologists, oceanographers, geochemists, and other geophysical researchers from all over the United States and Canada. What started as a question about summer ice in the Labrador Sea soon evolved into an investigation of remote-sensed passive microwave records of sea ice across the entire Arctic basin. What began as an inquiry about past ice behavior soon developed into a study of paleoclimatic records, ice drill cores, ancient iceberg armadas, and a worldwide system of deep ocean currents that was already being described by researchers as a key triggering mechanism for sudden episodes of global climate change. The "riddle of the ice" became a window into a larger and much more important question: the riddle of the Earth's climate.

The longer this investigation continued, the more I realized that I would need to sail *Brendan's Isle* north again—to witness the ice firsthand, to explore the questions it posed, and to *dramatize* this investigation, to rescue it from the computer screens and library carrels and scientific conference rooms where it had been taking place and move it back into the realm of concrete experience.

I was aware from the outset that people who are not sailors themselves might find it hard to understand why anyone would want to make such a voyage—especially in a conveyance as small and slow and uncomfortable as a sailboat. A sailboat is an anachronism in today's high-speed, high-tech world. Its rate of travel is unpredictable. It is inherently unstable and thus not much good as a platform for doing sophisticated scientific experiments. It is vulnerable to the elements—especially vulnerable to the ice—and under the extreme weather and sea conditions that one is likely to encounter in the Arctic, it is sometimes even dangerous.

As a sailor, I knew all these things. Yet I also knew that the story I wanted to tell would make sense only if I could somehow tell it within the context of a long sailing voyage. Part of this story was about a group of scientists struggling to understand the Earth's climate system by means of one of its most intriguing dimensions: the Arctic ice regime. But another part was about the ice itself—its complexity, its mystery, its massive scale, its indifference to all things human. It was this latter part that would require the voyage—for this part could only be told slowly, quietly, the way a sailor would tell it, as part of a broader experience of contact with the Earth.

A long sailing voyage is, by its very nature, an experience of contact with the Earth. But as generations of sailors have learned, it is an experience of contact that happens in a particular way. An anecdote about another, very different kind of voyage may help to illustrate this point.

A few years ago a wealthy acquaintance of mine who was also interested in ice and climate decided to take an eco-tour to the Canadian Arctic—not in a sailboat but in a nuclear icebreaker. The object of this voyage was to circle around a portion of the Canadian archipelago through the frozen Arctic Ocean—a feat that had never been accomplished. Along the way, the icebreaker's passengers were to be flown out over the frozen ocean in helicopters and transported across glacial ice fields in Skidoos to get a firsthand feel for the Arctic terrain.

When my friend returned from his journey, he showed me his photographs and told me excitedly of his adventures. Oddly, however, almost all his recollections were of the same kind of thing: the ship, the helicopters, the sophisticated electronic gear, the motorized sledges and Skidoos. He spoke with awe about the power of the mother ship as it crashed through three-meter-thick ice fields, and he bragged about the creature comforts available on board for passengers and crew.

Although the icebreaker never made it across the frozen Arctic Ocean as its operators had planned, this was nevertheless a voyage of conquest. And my friend came home not with an experience of the Earth or a deeper understanding of Arctic systems but with a renewed faith in the omnipotence of human technologies and in his own species' ability to control and manipulate nature.

How different it might have been if this traveler had journeyed into the ice in a sailboat, vulnerable to the natural forces around it, dependent for its very survival on its ability to bend to the forces of nature rather than to dominate and control them. The medium of the sailing voyage might have changed both what he saw and how he understood what he saw. Indeed, it might even have changed how he understood himself and all the rest of the human enterprise in the process.

The point should be clear: the next voyage I hoped to sail and the scientific story I wanted to tell were inextricably connected. Though not itself message, the voyage would nevertheless provide the medium that would color and inform everything that happened along the way. From the deck of a nuclear icebreaker, the story I'd learned might easily become a self-congratulatory celebration of the miracles of techo-science. From the deck of a little sailboat, it would become an investigation of vast and complex forces that science is only beginning to understand. It would become a story not of manipulation and control but of the vulnerability of our species and the honest search of scientists for the answers that might one day save us from ourselves.

Organized in the form of a sailor's log, *Riddle of the Ice* is an invitation for readers to enter imaginatively into the voyage and sail along. It is an opportunity for each of us to slow down, leave the confusions of the land behind, open our eyes and ears and minds, learn how to think like a sailor and to experience the Earth as a sailor might experience it. Then and only then will we be prepared to understand the efforts of scientists as they struggle to solve the riddle posed by the ice: not as self-important manipulators of nature but as sailors ourselves, as ones who have shared an experience of contact with the Earth in a posture of openness and with a spirit of humility, wonder, and awe.

# Riddle of the Ice

# Introduction

■

*August 4, 1991:* The afternoon was still and cold. The view from the summit of the hill I'd just climbed was fifty kilometers in every direction. To the west lay the rugged coast of southern Labrador, the steep shores of Chateau Bay, and the cove where the sailboat *Brendan's Isle* lay anchored near a grounded slab of sea ice. To the east stretched the gray surface of the Strait of Belle Isle, rimmed by the faint outline of Newfoundland's northern capes and covered, as far as the eye could see, with ice.

Icebergs, dozens of them, moved in haphazard patterns across the surface of the water. Floes of pack ice marched in broken rows down the axis of the swell. And to the north, just above the horizon, a ribbon of something not quite solid shimmered in the sky: ice blink.

I stepped to the seaward edge of the massif I had climbed and gazed out at the ice, trying to ignore the feeling of frustration that had been mounting in me for several days. Dammit, I thought—this just wasn't fair. I'd done my homework, studied the ice atlases and coast pilots, consulted with the ice forecasters. I'd even sailed along this coast twice in other summers—and I knew by every means available that these waters should have been free of sea ice weeks ago and that my crew and I should have had clear sailing to the north for hundreds of kilometers.

Yet the fact remained: we were stopped, surrounded by ice. Not trapped, exactly, for there was open water behind us and a clear means of retreat. But the plan I'd made so carefully last winter for sailing to the fiords of northern Labrador would have to be postponed now until another voyage, another year. All the work of preparing the boat, gathering crew, choosing routes, organizing rendezvous with other vessels—all was for naught. My crew and I had been waiting nearly four weeks for this ice to clear. Now we'd run out of time; it was too late in the summer to wait any longer.

As I stood at the edge of the precipice gazing out at the ice, I realized there was also something else that was bothering me—something far more important than my disgruntlement over our thwarted sailing plans. It had started with a remark that one of the ice forecasters had made during a telephone conversation a few days earlier—a speculation, really, that the massive influx of sea ice along this coast might somehow be connected to the warming of the Earth's atmosphere and that, as such, it might be a signal of much larger changes taking place in the climate system of the Arctic— possibly in the climate system of the entire planet.

Ironic, I thought, if this were so. Ironic if such a signal were to be happening here, at the ragged edge of the North American continent. Ironic, too, if it were to be happening like this, not in the heat of a fiery sun over a crowded city but in the solitude of an ocean full of . . .

"Oh there you are," a voice called out from behind me. "I've been looking all over for you."

". . . full of ice," I said out loud. Then I turned to watch Cherie, one of my shipmates, as she climbed across the rocky shelf toward me. She moved to the rim of the cliff and stopped, staring out at the sea in the same direction I'd been staring. Finally she spoke.

"You're thinking about what the forecaster from Ottawa told you about the ice, aren't you?"

I nodded, surprised at her ability to read my thoughts so easily.

"Could the forecaster be right?" she asked.

"He's certainly been right about the volume of ice out here."

"I mean could he be right about the *reason* for the ice. Could it actually be some kind of upside-down response to something we've done—we human beings—to cause the Earth's climate to start getting . . ."

"I don't know. I don't think anybody knows what 'we human beings' have been doing to the Earth's climate."

"But I've heard you say . . ."

"It doesn't matter what you've heard me say. I'm not a scientist. I'm a sailor. Whatever you've heard me say is just pure speculation."

I tried to change the subject. I gazed at the expanse of sea before us and gestured toward the north. "Look . . . do you see it? . . . That odd-looking shimmering just above the horizon?"

"It's some kind of mirage."

"A mirage . . . yes. It's called ice blink. It's the edge of the ice pack—or a reflection of it, anyhow—upside down in the sky."

Now it was my young shipmate's turn to stand and stare at the strange scene before us. A band of gray-white light floated like a fantastic island just above the rim of the sea. I waited in silence. It was nearly a minute before I spoke again.

"The fact is that neither of us knows why the ice is here this summer. It's possible that even the forecasters in Ottawa don't know why the ice is here. I'm feeling frustrated—just as you are—and I'd like to find an answer. I'm feeling disappointed, too, and angry at myself, because I keep thinking that somehow I should have been able to avoid this bottleneck we've sailed ourselves into . . ."

"Isn't there a chance we could still get through?"

I gestured again toward the north. "That's not just ice out there, it's a virtual *barrier* of ice, covering up to ninety percent of the sea in places. Ice so sharp it could slice through *Brendan*'s fiberglass hull like a knife through putty. Ice that begins maybe thirty miles north of here and blocks huge sections of the coast all the way to Baffin Island and beyond."

"I guess that means we're stuck."

"It means we have about thirty miles of navigable water ahead of us. Tomorrow we'll be able to sail north one more day—just to take a look. Then, next day, we'll be forced to turn around and begin the long journey home."

■

The ice my shipmate and I had seen from the cliffs overlooking Chateau Bay—the ice that had stopped *Brendan's Isle* some six hundred kilometers south of her intended destination—had remained throughout most of that region some four to six weeks longer than it should have, even in the worst ice year. Climatologists had started describing it as the most severe summer ice condition along the eastern coasts of Labrador and Newfoundland since the beginning of modern record keeping, nearly eighty years before.

The hardship of the people living on the coast had been extreme, especially in eastern Labrador. There are no roads in this part of the world—no means of traveling from one settlement to another except by boat or small aircraft. Emergency rations of food and medicine could be flown in by commercial airlift to settlements with airstrips. But all other supplies, all heavy foodstuffs, all machinery, all building materials, all motorized vehicles and the fuel to run them had to be transported along the coast each summer by the weekly freight boat.

As a result, the ice in July and August 1991 had shut down almost every activity that required heavy materials or fuel along a thousand kilometers of the Labrador coast. The ice had made a shambles of the inshore fishery, too, forcing boats to remain in harbor and destroying nets and traps. Everywhere on the coast the story had been the same: frustration and hardship. An entire summer lost. The worst ice year in any living person's memory. A disaster. A terrible catastrophe.

Ironically, as we'd sailed *Brendan's Isle* north from the east coast of the United States in early July, neither I nor anyone else on board had received the slightest warning about the ice we were about to encounter. On the contrary, we'd just come from a part of the world where a heat wave was going on—the fifth in as many summers. There was a drought, too, and another bad crop year. The grass had been turning brown in June, even before we'd left our home river on Maryland's eastern shore. The leaves on the oaks and locusts and maples along the riverbanks had been stunted and curling at the edges.

There had been talk in the newspapers and on TV about climate change, global warming. Most of this was media hype, I knew—weather talk. But in addition to all the speculation were other, more responsible voices. One of the loudest was that of global climate modeler Jim Hansen, director of NASA's Goddard Institute for Space Studies. Hansen had addressed a special committee of the U.S. Congress after the record-hot summer of 1988, citing the growing evidence of a worldwide atmospheric warm-up and urging policy makers to "stop waffling" on the subject of greenhouse-induced climate change. His warning was currently being repeated by growing numbers of scientists and amplified in nearly every form of popular media: the evidence was strong and mounting; global warming had started.

"Global warming, my ass," I'd muttered one afternoon as *Brendan's Isle* proceeded northward along the coast of Newfoundland, heading toward Chateau Bay. How could anybody be talking about global warming with an anomalous mass of sea ice out there blocking an area as long as the entire eastern seaboard of the United States?

As we proceeded north, I began calling the Ottawa forecast office of Ice Centre Environment Canada every day or two, whenever we were in port and near a telephone. Again and again the message from the Ice Centre was the same. The eastern coasts of Newfoundland and Labrador were both clogged with ice. Closed to shipping. The Strait of Belle Isle, too, was clogged with ice. Closed to shipping.

Why? I kept asking the forecasters. Why, in the midst of one of the hottest summers on record in North America, was the ice persisting here? And why in such prodigious quantity?

The answers I received always seemed to relate back to the randomness of nature. It was just one of those years, the forecasters seemed to be saying. There was ice weather, after all, the same as there was atmospheric weather. Both were anomalous, inscrutable, random-seeming. In spite of the contrast between the oppressive heat just a few hundred kilometers to the west and the barrier of ice less than a hundred kilometers to the east, we were, it seemed, the victims of an age-old truism: the unpredictability of nature.

On July 22 I made one final call to the Ice Centre. The

situation ahead, the forecaster assured me, was still not good. Granted, there were now a few leads of open water in the Strait of Belle Isle, thanks to several days of sunshine and strong west winds. But the coastal areas to the east were still jammed with pack ice.

In response to my inevitable question about why, the fellow I talked with that day began to explain an idea I had not heard before—something about warmer conditions in the high Arctic and how they might actually be a reason for the heavy ice.

I laughed, thinking he was making a joke.

"No, really," he insisted. "It's only hypothetical, of course. Nobody's been able to prove it yet. But the idea makes a certain amount of sense."

He went on to describe a series of ice events that may have taken place six months earlier in the dark of the Arctic winter somewhere far to the north. A lot of his explanation was difficult for me to follow, owing to my rather fuzzy grasp of Arctic geography at the time. But the thrust of his idea seemed to be that warmer than normal conditions in the vicinity of northern Greenland the previous winter had allowed a type of Arctic sea ice that should have been blocked by "bridges" of solid ice to continue flowing southward in the current, eventually invading the ice pack along the coasts of Baffin Island, Labrador, and Newfoundland.

"It's just the opposite of what you'd expect," he said. "Warmer conditions in the high Arctic. Channels of open water that would normally be frozen during the winter. Large, dense slabs of very cold Arctic sea ice flowing down through these channels, invading the softer, thinner ice farther south, aiding in its formation, postponing its melt. And finally, a couple of easterly gales during the spring to concentrate the ice and pile it up against the coast and— presto!—the worst ice condition in this part of the world in anyone's living memory. And not necessarily because the climate of the surrounding areas is getting colder but maybe—just maybe—because it's getting warmer."

■

Like most people in the industrial West, I've been no stranger over the past few years to talk about global climate change. And although I'm hardly a scientific expert, I've learned enough to be aware of at least one fundamental misunderstanding that many people seem to share.

"Greenhouse effect" and "greenhouse debate" are phrases familiar to almost everyone. But the phrases have been confused (occasionally, one suspects, purposefully confused) so that many people have been left with the impression that there is some kind of debate going on among scientists about the reality of the greenhouse effect.

There is no such debate. On the contrary, knowledge of the greenhouse effect is basic to scientific understanding of how the Earth's atmosphere works. As Harvard's Rotch Professor of Atmospheric Science, Michael McElroy, explains: "Were it not for clouds and gasses that trap heat emitted by the planet's surface—water vapor, carbon dioxide, and methane, for example—the Earth would be frozen over and life as we know it could not exist." What McElroy is saying is that the atmospheric mechanism we call the "greenhouse effect" is both necessary and ongoing. The phrase is merely a popular name for an age-old geophysical process of heat-trapping in the atmosphere that has kept the Earth warm enough for carbon life to evolve and prosper.

Just as there is no debate about the reality of the greenhouse effect, there is also no debate over the measurable increase in the concentration of greenhouse gasses in the Earth's atmosphere. Nor is there debate about the fact that much of this increase is tied to human activity—activity that has been taking place at an increasing rate ever since the beginning of the industrial revolution.

When Washington was inaugurated as U.S. president in 1789, for example, carbon dioxide ($CO_2$) accounted for about 280 parts per million (ppm) of the Earth's atmosphere, close to what it had been at the end of the ice age ten thousand years earlier. By Lincoln's time the proportion of $CO_2$ was 290 ppm. When Kennedy was inaugurated in 1961 it stood at 316 ppm. Today the $CO_2$ level has risen past 360 ppm and is continuing to increase at an accelerating rate as human population and the use of fossil fuels also increase.

The scientific debate, then, is neither about the reality of the greenhouse effect nor about the increase of greenhouse gasses in the atmosphere. The debate centers instead on amounts, timing, and long-term effects. *How long will it take and what concentrations of greenhouse gasses will be required to warm the Earth's climate system to particular levels? How fast will the warming take place? And what will the effects of this warming be on various essential human activities?*

All of these questions are subject to ongoing scientific dialogue—and in a few cases to serious disagreement. Yet in virtually every scientific prediction about changing climate, the bottom line remains the same. As long as our growing human numbers and our collective actions continue to add to the total concentration of greenhouse gasses, warming—and eventually dramatic warming—of the Earth's atmosphere appears inevitable.

It would be encouraging to imagine that our species might one day learn how to respond to clear and rational warning and to modify its collective behavior before a crisis occurs. But for some reason this isn't the way we seem to operate. The fact that our scientific experts can't yet point to irrefutable, human-induced changes in the climate serves for most as an assurance that we don't yet need to worry. In the absence of obvious and present danger, we seem willing to close our ears to the voices of caution and proceed with business as usual.

Given this tendency among people, many scientific watchdogs concerned with global climate warming have over the past few years started to search for concrete evidence of change. They are looking for what some have termed a "signal event"—an irrefutable occurrence in nature that will alter the way people think, removing the issue of climate warming from the realm of the theoretical and delivering it once and for all into the realm of the actual.

Surprisingly, there is a good deal of agreement about where to look for this sort of event. Most scientific predictions about future climate change are based on mathematical computer programs called General Circulation Models (GCMs) that have been developed to simulate the dynamics of Earth climate. Although these models are still far from unanimous regarding future climate scenarios, many indicate that the earliest and most dramatic changes will take place in the polar and subpolar regions, north and south.

The reason, quite simply, is the ice. When an area of land or ocean is ice-covered, that cover serves both as an insulator and as a solar reflector. As an insulator, it inhibits the transfer of heat between the land or water and the air. As a solar reflector, it has a high "albedo," or reflectivity, that directs solar radiation back into and through the atmosphere at a rate six to eight times greater than that of uncovered land or water.

What all this means is that the ice cover functions as a climatic accelerator. Growing ice cover increases cooling as it insulates the ground or water underneath it, reflects solar radiation away from itself, and encourages the growth of more ice. Shrinking ice cover increases warming as it exposes larger areas of the ground or water to the atmosphere, amplifies the absorption of solar radiation, and accelerates the decay of the remaining ice. Both sequences serve as positive feedback mechanisms that increase whatever changes may already be taking place in the climate system as a whole.

The signal event that climatologists are watching for—the event that, when it occurs, will confirm the onset of greenhouse-induced climate warming—is therefore very likely to be an ice event. But what kind of ice event? And what size will it be? And where will it take place? And when? And in response to what other forces?

■

The ice that *Brendan's Isle* encountered along the Labrador coast in August 1991 was both longer lasting and more extensive than any other in nearly a century. Could it be that this massive influx of coastal pack ice was working as some kind of reverse climatological barometer? Could it be that, instead of being an isolated event, it was actually part of a larger pattern of change across the entire Arctic? Could it even be that its reverse sign was obsuring what might otherwise stand out as a clear signal of climate warming?

These were questions I came home with in the fall of 1991. They were innocent questions, informed by the overwhelming presence of the ice and by an anonymous telephone voice speaking with the sometimes doubtful authority of government officialdom. But they were also questions that seemed important—questions that might lead to useful new insights about how the Earth works—questions that wouldn't go away until I'd found some answers.

The investigation that followed lasted three years. It was long and difficult, and I might have been tempted to abandon it except for a number of surprises that happened along the way.

The first surprise was the discovery that the entire geographical region between eastern Canada and western Greenland—the so-called Labrador Sea/Davis Strait/Baffin Bay (or LDB) area—was out of step with the predictions of virtually every General Circulation Model of Earth climate. According to the GCMs, the ice cover in the Arctic should have been receding and air temperatures should have been increasing as global atmospheric $CO_2$ also increased. Across most of the Arctic (and throughout the region when considered as a whole), this was indeed the case. But in the LDB area, just the opposite seemed to be happening. According to recent satellite and ground-station data, the LDB area had been functioning with an opposite sign at least since 1978, with the ice cover growing and mean air temperatures falling. According to earlier data, the LDB area had been out of step with the rest of the Arctic since at least 1965—and possibly since the beginning of this century.

The next surprise was that this climatological anomaly had a name: the "east Canadian/west Greenland cold spot." Along with its name, it also had a rather bothersome reputation among members of the scientific community. The growing sea ice season and the negative temperature signal of this region had become a problem for meteorologists, a stumbling block for ice forecasters, a fly in the ointment for global climate modelers who were looking for some sort of clear and unambiguous sign of change in the polar sea ice. In a phrase used by climate modeler Jim Hansen, "the problem of the east Canadian/west Greenland cold spot" was turning out to be a scientific toe-stubber, a "problem" for which no one seemed to have an answer.

The final surprise came not from climatologists who were looking at the present behavior of the LDB area, but from *paleo*climatologists—those who were looking at the area's past behavior. A pair of deep ice cores just then being drilled in the Greenland ice cap, a sequence of sediments deposited on the Atlantic Ocean floor during the last ice age, a record of sudden climate change inscribed deep in an Antarctic glacier, and numerous other ancient fingerprints around the Earth were all beginning to point to this area—to what had been happening for eons here—as holding the key to sudden climate changes in the geologic past and, by analogy, to similar changes that might take place in the future.

Without knowing it, *Brendan's Isle* and her crew and I had sailed our way into a scientific conundrum—a region that, although still imperfectly understood, was nevertheless at the center of a great deal of speculation and controversy. After returning home that fall and starting to communicate with researchers whose work was focused in this area, I came to realize that theirs was an area of inquiry whose data were new, whose questions pushed against the limits of scientific understanding, and whose answers were going to have far-reaching implications for every person living on the Earth.

■

There was no return to the ice for *Brendan's Isle* in 1992, and none in 1993. For one thing, the summer ice pack remained heavy along the eastern coasts of Newfoundland and Labrador in both these years—not as heavy as in 1991, to be sure, but heavy enough to hamper navigation both summers until late in the season. And for another thing, the reading and study that I'd been doing were nowhere near complete. For every question that had been answered by the experts, two more had arisen to take its place. For every conclusion drawn in the scientific literature, twice as many had been qualified or contradicted or postponed, pending further study.

Finally, however, in the fall of 1993, the decision was made: I chose my friend and longtime sailing associate Mike Auth to serve as mate, and together we spent one more winter finishing preparations on the boat and selecting a crew from a pool of more than a hundred candidates who had volunteered to take part in the journey. With the help of a sponsoring nonprofit organization, the Associated Scientists at Woods Hole, Inc., we solicited donated funds to cover some of the costs of the voyage. Then, in June 1994, *Brendan's Isle* once again pointed her bows toward northern Labrador—by way of Greenland this time—for a continued encounter with the riddle and a rendezvous with the ice.

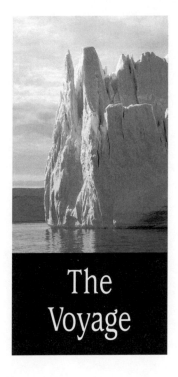

The
Voyage

# I.

# Northeast

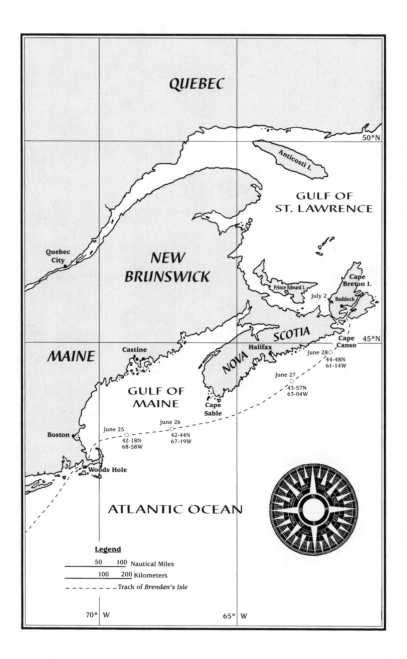

QUEBEC

50°N

Anticosti I.

GULF OF
ST. LAWRENCE

Quebec
City

NEW
BRUNSWICK

Prince Edward I.

Cape
Breton I.

July 2

Baddeck

SCOTIA

Cape
Canso

45°N

MAINE

Castine

Halifax

June 28
44-48N
61-14W

NOVA

June 27
43-57N
63-04W

GULF OF
MAINE

Cape
Sable

June 26
42-44N
67-19W

June 25

Boston

42-18N
68-58W

Woods Hole

ATLANTIC OCEAN

**Legend**

50     100  Nautical Miles

100     200  Kilometers

– – – – – – Track of *Brendan's Isle*

70° W

65° W

LOG: *Wind: south, 14 knots. Barometer: 30.06 steady. Weather: thin fog, broken overcast, full moon. Position: 55 miles NE of Cape Cod, Massachusetts.*

A hand touches me on the shoulder and nudges me into consciousness. "Skipper, wake up." Pete's voice. "You're on next. Are you awake?"

I open my eyes and find myself staring at a rectangular porthole above my bunk. The dim glow of moonlight enters through the window—the only light in the cabin. I turn slowly onto my back and feel the rhythm of a long, regular swell lifting and falling underneath the boat. I listen to the hissing of water moving fast along the hull next to my ear.

"Yes, yes . . . I'm awake. Thanks," I say to Pete. "What time is it?"

"Ten minutes to twelve," Pete says, stepping back into the galley. "It's a beautiful night out there."

I lie still for nearly a minute after Pete has left the cabin. I need to retreat from the image of whatever dream he has just interrupted. I need to let the pieces of my consciousness organize themselves into a coherent whole.

Ten minutes to twelve. The change of watch. *Brendan's*

*Isle* is sixteen hours out of Woods Hole, Massachusetts, somewhere north of George's Bank in the Gulf of Maine. She's sailing northeast with the wind on the quarter, the liquid miles rolling effortlessly under her keel.

I sit up, swing my legs over the edge of the bunk, drop slowly to the floor. I'm already dressed in blue jeans and a turtleneck shirt and sweater—the air is cool tonight and I never bothered to take them off when I left the deck four hours ago. In the darkness I fumble around until I find the black rubber sea boots and the wool cap and the nylon safety harness and tether that I'd left together on the floor at the foot of the bunk.

My eyes have adjusted to the moonlight well enough now that I don't need to use the overhead electric lamp. I know it's better to leave it off and finish dressing in the dark. I'll need my night vision, anyhow, in a few more minutes up on deck.

Pete is standing in the galley next to the stove when I emerge from the aft cabin. "Coffee's in the thermos," he says. "Hot water's almost ready for tea."

The light from a single kerosene lamp flickers against the wall on the far side of the dinette table in the main saloon. Underneath it I can just see the dim outline of my watchmate, Amanda. She pulls a sweater down over her head and shoulders, then looks at me with sleepy eyes.

"Hi," she says, in a hoarse whisper.

"Good morning," I say. "Did you sleep all right?"

She nods, staring at her knees. She is trying to pull a pair of jeans up over her long underwear without standing up. She is obviously still in a half stupor as she struggles in the darkness with her belt and zipper.

"Take your time," I say. "I'm wide awake. Why don't you just stay down here, have a coffee, watch the radar for a while. I'll take the first hour on deck."

She looks at me and nods again. "Okay, sure," she says. "Thanks."

In the cockpit I clip the tether of my safety harness onto one of the two long Dacron jack lines that have been rigged along either side of the deck. I slide aft and duck around Mike as we exchange places at the helm. Our tethers become fouled, and I have to unclip for a few seconds to let him pass.

There is no need for the safety gear tonight. The seas are gentle and the boat is well off the wind. We're wearing the safety gear so that it will become a habit—an unquestioned part of the routine. We're wearing it for all the other nights that are sure to come when the seas won't be so gentle and the wind won't be so kind. Mike and I and the others have accepted this fact. None of us complains about the inconvenience.

After Mike has left the deck, I stand at the steering station, trying to orient myself to the night. The full moon lights the sky in a diffused glow. A thin veil of fog obscures the horizon so that there is no perception of distance, just a cottony film of silver light that envelops the boat. But there are hundreds of miles of open ocean ahead of us, with no weather to worry about, no coastal shipping lanes to cross. I know that with Amanda checking the radar from time to time below, all I need to do is steer. So I hunker down in the seat behind the wheel, listening to the sound of water trailing off the stern, thinking about where in the world we've come from, where we are trying to go.

Twelve hours earlier, as we sailed past Race Point at the end of Cape Cod, a strange thing happened—the same thing that always seems to happen when you leave the land in a little sailboat. As we skirted around the Cape, the beach was only a half mile away and the details of hills and towers, buildings and roads, automobiles and clusters of people stood out in vivid relief. Then in a few more minutes *Brendan* found the south wind again, and as she pulled away from the land, the details began to blur. At two miles off there were no people any longer. At three miles the roads and automobiles were gone. At five miles the buildings and the beaches disappeared. At seven there was only the gray-green silhouette of hills at the rim of the horizon, punctuated by the tops of towers. At ten the last remnant of the North American continent dropped unceremoniously into the sea, and we were alone.

There is an odd feeling that comes at such a moment. As the land disappears, you feel yourself growing less distinct— as if the mirror that you normally see yourself in has suddenly started returning a weaker reflection. You begin to think about scale—about how small the land is and how large the sea. You begin to think about the Earth as a water planet—with over seventy percent of its surface covered with this liquid wilderness and only thirty percent solid and dry enough for human habitation. You begin to think about all the petty strutting and preening that goes on back there, and about how quickly it disappears beneath the horizon. Inevitably, you also begin to think about how small and unimportant you must be—and perhaps how small and unimportant the entire human enterprise.

Tonight you are a sailor, moving along with the forces of nature rather than against them. You are a minion of the wind—only partially in control of where you go—just as in a few more weeks you'll be a minion of the ice, forced to do its bidding rather than your own. Maybe this is another reason why you feel so small tonight and so disengaged from the commotion and noise that you've so recently been a part of ashore.

Joseph Conrad once wrote, "The true peace of God begins at any spot a thousand miles from the nearest land." You're not a thousand miles from the land tonight—you're not even a hundred. But the feeling that Conrad was talking about must be the same as the feeling you have here. The feeling of being safely at sea with the land and all its confusions well astern. The feeling of sailing fast, listening to the odd, nighttime chatter of seabirds, watching the moonlight illuminate the tops of waves, moving your body in time with the rhythms of the boat. The feeling of connecting.

*June 26, Sunday morning. 42/26N, 68/09W.*

*LOG: Wind: southeast, 8 knots. Barometer: 30.10 steady. Weather: hazy sun. Position: Gulf of Maine, 50 miles N of Cultivator Shoal.*

The voyage we are making this summer actually started a week ago with a passage from the Chesapeake Bay to New England, followed by several days' layover in Woods Hole, Massachusetts, where we met the last of the four crew members to join the boat, Amanda. Yesterday as we set out across the Gulf of Maine toward Nova Scotia, each one worked to adjust to the rhythms of life under sail. Today, the first full day at sea, we begin the difficult task of learning how to sail together as a crew.

Blue and Pete, the two youngest members of the group, have been trying hard during the past week to learn the routines of the vessel and to figure out what is expected of them. Mike and Amanda, on the other hand, have both been to sea aboard *Brendan's Isle* many times before. Mike has sailed nearly ten thousand sea miles with me—including a high-latitude passage across the North Atlantic in 1984. Amanda sailed with me first as a student aboard my old sail-training schooner *Dawn Treader* when she was in high school.

Several years later, in the summer of 1988, she served as a deckhand aboard *Brendan's Isle* during a voyage to Labrador. Like Mike, she simply needs time to adjust again to shipboard routines.

Blue is the first of the new crew members that I've gotten to know. We were watch partners for the passage from the Chesapeake Bay and were thus able to share much of our deck time together. (Blue's real name is Michael Browne, but with two other Mikes on board this summer, Mike Auth and I, I've asked him to come up with another nickname—and "Blue" was his immediate choice.)

Blue is twenty-two years old, just graduated from Brown University a few weeks ago with a degree in mechanical engineering. He's had no real-world experience yet—hasn't had time. But he's had a few summer jobs, all of them working around boats. Two years ago he was a deckhand aboard a Maine sail-training schooner, *Ocean Star*. Last summer he worked out of Sitka, Alaska, aboard a rockfish dragger.

In many ways Blue seems a bundle of contradictions. His mop of curly red hair, his stocky, athletic build, his huge shoulders and arms, his habit of laughing awkwardly whenever he feels self-conscious, all make him seem almost oafish at times. Then he opens his mouth to ask a question or remark on an idea he's been thinking about, and it's suddenly obvious how extraordinarily bright he is. He has a memory like a steel trap, and he learns faster than anyone I've ever sailed with.

Blue is also something of a sailing junkie—the same as me—although we've already had our first difference of opinion. He is a purist, it seems, and doesn't want me to use the engine—*ever*. Several days ago, when the wind dropped and came ahead, I asked him to fire up the diesel while I stowed the headsail. I tried to explain that we had promised to meet our sponsors in Woods Hole for an official launch of this expedition in less than two days' time.

"We shouldn't have to worry about deadlines," Blue said.

"Maybe so. But we *do* have to worry about deadlines," I said. Then I told him again to fire up the engine, and he stared back at me with a look designed to kill. He pouted during the rest of our watch like a little kid who'd just been told he couldn't eat any more candy.

Blue's bunkmate, Pete Johanssen, is two years older than Blue, making him the senior member of the twosome. Pete is quieter and more introspective than Blue—less dramatic, less in need of attention. This summer Pete has brought a banjo with him. Whenever he finds a private moment, he sits alone on the deckhouse roof and picks out the single notes of an old folk song. He gazes off at the horizon, his curly dark hair rustling against the collar of his T-shirt, his chiseled features set in an expression of almost perfect calm, and he seems transported to some other place and time.

Pete is a scientist by temperament as well as by schooling. He was trained at Williams College as a geologist, and he's spent part of the past year on a school ship in Stamford, Connecticut, teaching young children about estuarian ecology and the other part of the year working as a lab assistant on an oceanographic research tug, the *Maurice Ewing*. Last spring he asked me if he could bring a few scientific instruments along with him to Greenland so that he could make some simple measurements and keep a scientific journal of the voyage. I gave him permission—along with a reminder about the scarcity of stowage space on board—and then I promptly forgot about his plan.

It was only after he arrived in Maryland several days ago, balancing several extra boxes of scientific gear on top of an already full complement of boots and bedding and duffle bags, that I began to worry about where he was going to fit it all. There were thermometers, bottom dredges, sample bottles, a plankton net, a microscope, test tubes, petri dishes—so much paraphernalia that he finally had to stow the overflow in his bunk. I found myself wondering if he'd be able to find any room in there afterward to sleep.

My old friend and shipmate Mike Auth is captain of the second watch, so I don't get much time to see him during an offshore passage like the one we're doing now. But there's no need, anyhow, for introductions between Mike and me. We've sailed together for fifteen years and thousands of miles. Last fall, after he'd made a firm commitment to do the Greenland voyage, he set to working as hard as I was to help get the boat ready, interview prospective crew members, and meet with scientific researchers and project advisers.

Mike's thick white hair and handlebar mustache lend him a dignified, almost military air. But he speaks quietly, with just the hint of a drawl, and he exudes a kind of down-home charm that immediately puts people at ease. He is a good shipmate—and what's just as important, he is a good communicator. He is more relaxed than I am, more patient with people. I will count on him this summer to serve as an intermediary between me and the younger members of the crew—to translate my expectations to them, to let them know that I will always be harder on myself than I will sometimes seem to be on them.

The final member of the crew is the one who just arrived in Woods Hole—Amanda Lake. I missed Amanda's company during the first leg of this voyage, missed her quiet self-assurance and her zany sense of humor. I've sailed with her, as I've said, many times before. Yet until a few days ago, I hadn't seen her for nearly two years, since she left her job at the National Audubon Society in Washington, D.C., to go back to graduate school.

I know that Amanda is a fine sailor—sensible, resilient, confident of her own abilities. I also know that she's a person who understands intuitively how to keep things working smoothly in a group. The only thing I don't yet know is how she'll get along—as the only female on board—with her new shipmates. But I have a feeling it won't take long today before we begin to find out. . . .

The air begins to warm, the haze lifts, and the breeze fills in from the southeast. I know that *Brendan* will move better with some extra sail forward, and I decide that it may be time this morning for a little foredeck drill. I stand behind the helm; Mike waits at the mainsheet winch. The others move forward, and I watch to see what will happen next.

"Staysail," I holler. "Let's get her up."

Blue steps to the mast, uncleats the staysail halyard, and wraps a turn around the barrel of the winch. Then Amanda steps up behind him and taps him on the shoulder. She is holding the winch handle in her hand—it seems she grabbed it from its holster at the base of the mast as she came forward.

"Looking for this?" she says, winking at Pete and waving the handle just out of Blue's reach. "How about it, Blue? Mind if I give that halyard a try?"

I smile at the look of surprise on Blue's face, and I think to myself, Yes, that's my old pal Amanda. She'll be able to handle these guys just fine.

*June 26, Sunday afternoon. 42/44N, 67/19W.*

*LOG: Wind: southeast, 12 knots. Barometer: 30.12 rising. Weather: clear. Position: Brown's Bank, 49 miles SSW of Cape Sable, Nova Scotia.*

As often happens on a fast, easy passage, the combination of gentle seas and a sunny afternoon finds everyone rested and awake at the same time. Blue and Amanda and Pete are in the cockpit, talking quietly, sharing the work of steering. Mike is on the foredeck reading, while I spend the afternoon down at the navigation station engaged in my first serious session with the Canadian Ice Centre's radio-fax ice reports.

Yesterday I programmed a series of transmission times into the fax receiver so that by noon today a complete sequence of ice maps—from Baffin Bay all the way down to Newfoundland—has been automatically printed out and is waiting in a jumble underneath the machine. Now I cut the maps into appropriate sizes and arrange them on the table in front of me.

There was a time not long ago when maps like these wouldn't have made much sense to me, for in order to understand such maps, you need first to know something about the various types of ice that occur in nature—their names, their characteristics, where each one is formed, and how each behaves. There is no single source for this kind of information. Some of it can be found in the ice atlases and other standard publications of the two national ice services. Some of it can be found as you explore the scientific literature or talk with scientists who are studying the ice. And some of it can be found only as you actually journey to the ice and learn—by the seat of your pants—how it works.

The maps on the table before me contain useful and timely data about three basic types of ice. More than ninety percent of this ice is known as sea (or pack) ice. A much smaller amount is known as landfast (or fast) ice. And a smaller amount yet—noted on the maps only by means of a single generalized symbol (an upside down $V$)—is known as iceberg ice.

As its name suggests, sea ice is a creature of the open ocean. Composed entirely of saltwater, it forms when the surface of the ocean cools below about −1.8 degrees Celsius (about 28 degrees Fahrenheit). It forms in "floes" of various sizes and shapes and never attaches itself permanently to the land, so that it's always moving, always being transported from one location to another by the winds and currents.

Two subcategories of sea ice are important to know about: first-year ice and multiyear ice. First-year ice is just what it sounds like: sea ice that forms in the autumn or early winter of one year and lasts only one season, melting before the following autumn. This type of ice is usually white, owing partly to the salt brine and air bubbles trapped inside during its formation. It grows to various thicknesses, usually between about a third of a meter and two meters (although it can sometimes multiply this thickness several times by "rafting" when forced together with other ice under pressure).

The other subcategory of sea ice, multiyear ice, is defined by NASA scientists as "ice that has survived at least one summer melt period," so that it reenters the winter ice regime the following autumn. Multiyear ice is something of a metamorphic oddity, since it actually changes its chemical structure as it survives through its first year of life. During the comparatively warm summer months, meltwater collects on its upper surfaces and percolates down through it, causing desalinization of the surface and underlying layers and eventually carrying the salt into the seawater below. When the ice refreezes the next winter, it takes on a bluish color and a rough, hummocky appearance, it becomes denser and colder, and it turns almost completely fresh to the taste.

The point to keep in mind about *all* sea ice—the point I must think about as I try to understand these ice maps—is that it moves. It moves with the current. It moves with the wind. It moves at times as fast as a small boat can move. And what this means is that these maps, as useful as they are, are really lies. They present the ice as static when in fact it is just the opposite. They tell you where the ice is now, when common sense should tell you that it's already someplace else.

The only kind of ice that doesn't move is the kind called landfast (or fast) ice. This is the ice that people who live near the seacoast are most familiar with—the ice that grows connected to shorelines, trapped in bays and sounds. Landfast ice is usually grayer than sea ice, also smoother and less porous. Because it tends to grow in shallower water it's often the first ice to form in the autumn and the first ice to melt in the spring. It can sometimes be uprooted from its anchoring points along the shore by a combination of high tides and strong offshore winds—and then it joins the sea ice and becomes part of the moving pack. But most often it just stays where it was formed, waiting for the longer days and warmer temperatures of spring and summer to melt it in place.

The maps that I've just received show that there is still

extensive landfast ice in the inlets and fiords of both northern Labrador and western Greenland. One of the reasons
*Brendan* waited until late June to leave the U.S. coast, in fact,
is the presence of this ice. There's no point in pushing north
too quickly on a voyage such as this, for landfast ice has a
natural timetable, and if we arrive too early we'll just have to
wait for it to disperse.

The last type of ice depicted on these maps is the icebergs.
Unlike sea ice, icebergs are composed not of liquid water but
of compressed snow. Formed over the land as glacial ice
sheets, they are the result of snowfall that has piled up over
thousands of years, eventually compacting under its own
weight into a pure, white, freshwater ice whose crystalline
structure contains billions of trapped compressed-air bubbles—the residues of ancient atmospheres.

The increasing weight of the snow eventually causes the
ice sheets to begin to move. Under the force of gravity, they
slide along the rock beneath them until slowly, inexorably,
they find their way to the sea. Here they begin to "calve"
into huge, fantastic slabs—icebergs—that tumble into the
water. In the places where the water is deep enough, the
icebergs float away, driven by the current and the wind to
become, in the words of one of the Canadian ice atlases,
"immense, moving local hazards to navigation."

Icebergs migrate along the Labrador and Baffin and
Greenland shores in huge numbers: estimates of iceberg production in Baffin Bay alone run as high as forty thousand per
year. During the winter they become frozen into the surrounding pack ice so that they move with it, responding as
part of a unified field to the forces of wind and current.
Icebergs look like vast ice mountains when they are frozen
into the pack this way—jagged peaks and cliffs, massive slop-
ing inclines, huge tabular mesas of searing white—the
purest, whitest white that happens anywhere in nature.

In the spring, when the pack begins to break up, the icebergs are released and allowed to move on their own—and here is where their unique character becomes apparent. The pack ice, lying as it does on the surface of the ocean in horizontal slabs, responds as much to the wind as to the current. But the deep-drafted icebergs, with up to ninety percent of their mass underwater, seem almost impervious to the wind; they move, instead, in the direction the water moves. On days when the wind is blowing one way and the current is flowing another, the bergs can actually appear to be smashing through the sea ice like huge white ships, scattering the floes aside and leaving a trail of open water astern.

In summer, when the bergs are also subject to melt, they begin to fracture and roll and break apart, creating clusters of smaller iceberg shapes: "bergy bits," miniature bergs the size of small buildings, rising four to six meters above sea level and covering areas of several hundred square meters; and "growlers," much smaller slabs floating horizontally in the water and protruding less than a meter above the waves.

Icebergs are almost always visible on radar, so they pose relatively little danger to a boat like *Brendan's Isle,* even in the fog or at night. Bergy bits are also usually easy to see—if not always by radar, then at least by a careful watchkeeper posted on the bows. Growlers, on the other hand, are usually too low and too widely scattered to see, either electronically or by eye, until they are very close. These small iceberg pieces can thus pose some of the greatest danger of any ice to a small boat; yet, like icebergs, they are never charted individually on the ice reports but are designated only in large, general areas with the upside-down *V* and the phrase "bergy water."

The ice maps spread before me this morning contain almost as many questions as answers. They show where the main body of sea ice is located now—but they provide almost no indication as to where it may move as we travel north. They show two potential routes to Greenland—one passing to the north of Newfoundland, the other to the south—but they don't yet indicate which route *Brendan* will be able to sail. They show large areas of bergy water along both routes, but they don't say anything about the size or concentration of the bergs within these areas or about the frequency of smaller iceberg pieces that may pose the greatest danger to *Brendan*'s fiberglass hull.

Most of these questions, I realize, will just have to wait. There are too many forces at work in the ice—too many variables in the system as a whole—to permit anyone to predict this far in advance exactly what the ice is going to do.

There does appear to be one bit of good news, however, something that I've been noticing on these reports since about the middle of May. I pick up a chart of the Newfoundland coast and the Strait of Belle Isle and stare at a line representing the western edge of the ice pack. There . . . it's still there . . . along the coast . . . an area of open water between the shore and the ice. Every time I've looked at these charts for the last month, the opening has grown wider. What I'm beginning to suspect is that there may have been a period of strong west wind up there helping to blow the ice out to sea.

I wince at this thought, knowing it's still too soon to tell, realizing that I shouldn't jinx this voyage with such an optimistic forecast quite so soon.

"We'll see," I say out loud. "Maybe we'll get lucky this time—if the wind doesn't turn east again and blow it all back ashore . . . maybe we'll get lucky. . . ."

LOG: *Wind: southwest, 16 knots. Barometer: 30.11 falling. Weather: overcast, haze. Position: 39 miles SSE of Chebucto Head, Nova Scotia.*

A crackling noise fills the cabins this evening, competing with all the other noises *Brendan* makes as she rolls down following seas. I am seated in front of the single-sideband radio, trying to make contact with a schooner called *Bowdoin* in her home port of Castine, Maine. I've promised her skipper that I'd try to call as soon as *Brendan* turned the corner at Cape Sable and started sailing northeast along the coast of Nova Scotia. This call is a test run for a radio contact that we both hope to continue on a regular basis for the rest of the summer.

*Bowdoin,* I should emphasize, is not just any schooner—and this radio contact, if it works, is not for the purpose of idle conversation. The schooner is currently a sail-training ship for the Maine Maritime Academy in Castine, but she's better known for her historical role as an Arctic explorer and research vessel, the most heavily campaigned American Arctic expeditionary vessel of the early part of this century. Originally launched in 1921, she is a twin-masted gaff-rigged schooner, eighty-nine feet long, heavily constructed of white oak planking and greenheart sheathing over double-sawn oak frames. Her designer and original owner was the American explorer, scientist, and educator Rear Admiral Donald MacMillan. As a young man, MacMillan had been a member of Admiral Peary's historic first expedition to the North Pole in 1909. Later, convinced of the scientific and educational importance of further exploration of the eastern Arctic, MacMillan designed his schooner as a state-of-the-art research vessel for the express purpose of returning to the coasts he had visited with Peary. Thereafter Captain Mac, as he was affectionately known, campaigned *Bowdoin* for twenty-six voyages between 1921 and 1954 to the coasts of western Greenland and eastern Canada, following the same route that *Brendan* will follow this summer.

After MacMillan's death in 1970, *Bowdoin* languished for nearly a decade until a major restoration project rendered her once again fit for sea. Soon afterward she was donated to the Maine Maritime Academy, where she's been campaigned ever since as an Arctic sail-trainer, following the traditions that MacMillan began. In 1991 (the same year *Brendan* was stopped in the ice near the Strait of Belle Isle), *Bowdoin* sailed to Greenland and northern Labrador by way of Saint John's, Newfoundland, under the command of Captain Andy Chase. This year she plans to sail to Greenland again, departing from Castine in a few days with a crew of sail-trainees under the command of Captain Elliot Rappaport.


The radio crackles and hums as I dial around the six-megahertz band, looking for a channel that's clear of traffic. As soon as I've found one that will work, I flip back to the calling channel.

"*Bowdoin, Bowdoin, Bowdoin,* this is sailing vessel *Brendan's*

*Isle, Brendan's Isle,* Whiskey, Romeo, Charlie three-six three-six . . ."

Almost immediately I hear Elliot's voice intoning *Bowdoin*'s name and call sign. We move to a working channel, exchange greetings and positions, and turn to the subject of itineraries.

"Are you still planning to leave Castine on the first of July?" I ask Elliot. "Will you be sailing around southern Newfoundland by way of Saint John's?"

"Affirmative, affirmative," says *Bowdoin*'s skipper. "And what about you, *Brendan's Isle?* Will we see you in Saint John's on your way north this year?"

"It's still hard to say, hard to say. The ice looks lighter than in ninety-one, that's sure. But I'm going to wait to make a final decision on the Saint John's route until just before we leave Cape Breton."

"Roger, roger. Sounds like a good plan."

We exchange a few more items of news, then arrange to make another contact as soon as *Bowdoin* begins her passage across the Gulf of Maine. I sign off and shut down the radio. Then I turn and find Pete standing a few feet away, staring at the radio and working his mouth in a kind of half stutter.

He speaks haltingly, as if he's still trying to think through what he wants to say. "I don't . . . well exactly understand . . . you know . . . if the ice is still in the north . . . then how . . . ?"

Eventually it becomes clear that he still can't understand our strategy of having two alternative sailing routes to Greenland: one through the Strait of Belle Isle to the north of Newfoundland, the other out past Saint John's and across the Grand Bank's to the south of Newfoundland.

What's the point of trying to detour *south* around the ice, Pete wants to know. Doesn't the spring pack ice advance down through the Davis Strait like a barrier, moving with the prevailing winds and currents? And if that's so, then won't it still be there, waiting for us, as soon as we start sailing north again?

Good question, I say to my shipmate—in fact, a *very* good question. I begin to wonder how many others on this boat may also be asking themselves the same question.

"How about brewing up a thermos of coffee," I say to Pete, "while I invite anybody who's interested to join us around the dinette table for a little geography lesson."


The answer to Pete's question goes back a long way. It goes back to *Brendan*'s first contact with the schooner *Bowdoin* in 1991. Then it goes back some four hundred years earlier—to the voyages of the English navigator and master mariner John Davis, the first modern European to successfully circumnavigate this region.

Everyone aboard *Brendan* knows about our ill-fated dash to the Strait of Belle Isle three years ago and our inability to find a way through the ice. What some may not know is that *Bowdoin*'s skipper and I had made a plan that year—as we have again this year—to rendezvous in some remote anchorage far to the north, to visit for an evening and swap stories.

*Bowdoin*'s itinerary in 1991 was to sail out to Saint John's, at the far southeastern corner of Newfoundland, then move north to the midcoast of Labrador as far as the ice would permit before heading across Davis Strait to Greenland. Meanwhile, I was planning to sail the "shortcut" route to Labrador—up the Gulf of Saint Lawrence and through the Strait of Belle Isle—which meant that we could plan our rendezvous somewhere along the Labrador coast.

But while *Brendan* was dragging her heels in western Newfoundland, waiting for the ice to clear farther north, *Bowdoin* was modifying her plan, sailing east across the Grand

Banks, then turning north and sailing in ice-free waters up the center of the Labrador Sea toward Greenland. While *Brendan* was feeling her way along the edge of the ice pack near Chateau Bay, Labrador, *Bowdoin* was sailing in the midnight sun across Disko Bay, Greenland, four hundred kilometers north of the Arctic Circle. While *Brendan* was retreating down the Gulf of Saint Lawrence with her tail between her legs, *Bowdoin* was following the first icebreaker of the summer into Nain, on her way back down the Labrador coast.

Incredibly, *Bowdoin* had managed to accomplish her entire summer itinerary *during the worst ice year in this century along the Labrador and Newfoundland coasts*. The reason for her success had to do with a circular sailing route that followed the prevailing currents around the LDB area—a route that had been pioneered four hundred years earlier by English navigator John Davis.

A look at a modern current chart of the LDB area explains the logic behind the so-called Davis route and illustrates the behavior of the sea ice in this region. Researchers know that the major surface currents here flow in a large, counterclockwise circle around the perimeter of the area. Within this larger circle are several secondary circles (or "gyres"), both large and small, all flowing in the same counterclockwise direction.

Two primary sources feed this system: a warm-water source to the south—the Gulf Stream—and a cold-water source to the north—the Baffin Bay side channels of Smith, Jones, and Lancaster Sounds. Currents on the east side of the system flow northward from the Gulf Stream, transporting warm water to the Greenland coast. Currents on the west side of the system flow southward from the Arctic Ocean, transporting cold water to the coasts of Baffin Island, Labrador, and eastern Newfoundland.

What all this means is that the sea ice in this region follows

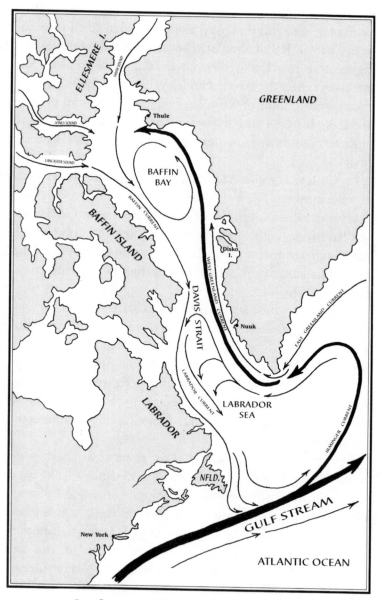

**Surface currents in the LDB area.**

two distinct patterns of behavior, one for the cold-water side along the Baffin/Labrador coast, the other for the warm-water side along the Greenland coast. Along the Baffin/Labrador coast, the sea ice season is always longest and the ice cover is always heaviest, whereas along the Greenland coast the sea ice forms late and disperses early, and there is often a coastal lead of ice-free water all the way to the Arctic Circle and beyond.

Primarily because of this cold/warm contrast in the temperature of the ocean surface, the sea ice virtually never forms an impassable barrier across this region. Instead, beginning somewhere on the mid-Baffin coast, the ice forms exclusively along the western (cold-water) side of the region—and it *stays* there—so that the only barrier that is ever created is the one along the coasts of Labrador and Newfoundland.

This pattern has been well documented by mariners over the centuries, photographed from surveillance aircraft since the early 1940s, and, most recently, measured by orbiting satellites from hundreds of miles above the Earth's surface. The real mystery is that in a time long before any of this documentation was available—indeed, in a time long before any charts were available—a European sailor in command of three small wooden ships was able to double Cape Farewell, sail to the Arctic Circle and beyond, and make his way back down the coasts of southern Baffin Island and northern Labrador in a perfect rendition of the counterclockwise circle route. John Davis did just this in 1585—and again in 1586 and 1587. As Columbus had done almost a century earlier, Davis pioneered a difficult circular sailing route on his first attempt—a route that has never been improved upon.

Davis obviously had a kind of sea sense, a knack for sensing patterns and going with the flow. Some are born with this sense, while others have to learn it. I'd always thought I had fairly good sea sense. Always thought so, that is, until the summer of 1991 when *Brendan* got bogged down in the ice in southern Labrador.

"If you study the ice atlases, you'll find out that there's only one way to get completely stuck," I say to my shipmates, "even in the worst ice year. And that one way is to sail up to the Strait of Belle Isle—as I did—while the pack ice is still heavy along that coast.

"Don't get me wrong—there's always a strong temptation to use this route. For one thing, it saves three hundred sea miles in overall distance. And for another thing, it puts the prevailing winds and seas behind you instead of dead on your nose."

"What about this year?" asks Pete. "Will we be trying again to get through the Strait of Belle Isle?"

"I don't know, Pete. It's still too soon to tell. I'm not quite as smart as old John Davis. But I promise you right now, unless the strait is wide open, I won't make the same mistake again. If we're able to get through the ice, fine. If not, we'll sail this boat *all the way to Europe and back* to make sure we get around it!"

*June 28, Tuesday. 44/48N, 61/14W.*

*LOG: Wind: nil. Barometer: 29.96 falling. Weather: thick fog, visibility 300 yards. Position: 30 miles SW of Cape Canso, Nova Scotia.*

Last evening less than an hour after I'd finished the radio contact with *Bowdoin, Brendan* slowed to under three knots and started slatting and rolling, throwing the wind out of the sails. The fog settled down heavy after dark, the mast and booms dripped like rain, and I decided it was time to turn on the engine.

Now, fifteen hours later, *Brendan* is still in thick fog, still under power, and what before had seemed a responsive "minion of the wind" has been transformed into something loud and dull and plodding—an engine-powered craft droning its monotonous way across glassy seas.

I, for one, have become almost catatonic. The hours drag on, and the boat is filled with engine roar. Nuances of color and sound and movement disintegrate. Conversation stops. The feeling of connectedness that only a day earlier had seemed so all-pervasive suddenly dissolves. The passage begins to feel like a task, to be gotten over as quickly as possible.

There seemed two good reasons last night for continuing under power: the batteries needed charging after three days of sailing. And the forecast for this afternoon was for strong south wind, promising a fast finish and daylight approach to Saint Peter's Bay on the south coast of Cape Breton Island.

The forecast, however, was wrong. By midafternoon there is still no wind. *Brendan* drones on, and I grow cranky and irritable.

Just after the change of watch this afternoon, I find myself sitting in the cockpit, hunched over the steering wheel, staring into a horizonless sky. Blue climbs up on deck and squats across from me, looking about as miserable as I feel. I have an idea I know what he's going to say.

"Isn't there something we can do to turn off this engine?"

"Sure there's something we can do. We can take it out of gear and switch off the key. But we're not going to do that."

Blue gives me his best Captain Bligh scowl, curls his lip in an exaggerated pout, and begins to fidget with his hands. "How can you call yourself an environmentalist?" he says. "How can you keep talking about 'climate signals' and 'greenhouse gasses' and 'global warming' . . . and then just stop sailing and run this diesel engine for fifteen hours straight? What about the oil that seeps into the bilges and gets pumped overboard? What about the $CO_2$ emissions? . . . Twenty-two pounds added to the atmosphere for every gallon of fuel we burn . . ."

Oh great, I think. *Great.* This is just what I need—a sermon. A goddamn *sermon* from an insubordinate, holier-than-thou twenty-two-year-old child prodigy—and one of my own crew members at that!

I stare back at Blue with *my* best Captain Bligh scowl. "I don't know whether this makes any difference to you, Blue, but we've still got more than five thousand miles to sail this summer. And we've got something like eighty-three days to sail them in. You want to do the arithmetic? Go ahead. Subtract the days we're planning to be in port. Subtract the days we'll be stopped because of weather or ice. Divide what's left by the number of miles we still need to travel—and you've got some idea of the magnitude of what we're trying to accomplish."

Blue looks as though he wants to say something in reply. Then he seems to change his mind. Maybe he senses my mood. Maybe he realizes it's not a good time right now for an argument. Or maybe he just gives up on me, figuring I'm too much of a hypocrite, too far gone for help.

He turns without another word and heads back down into the cabin. I return once more to listening to the drone of the engine, staring out into the fog.

Damn fellow is just being obstinate, I think—just trying to get my goat. Well, he can complain all he wants, but it's not going to do him any good. We can't sail when there's no wind. And the numbers speak for themselves. We're trying to take a fifty-foot sailboat to goddamn *Greenland* this summer! Blue just has no idea yet what that means.

*July 2, Saturday evening. Government Wharf, Baddeck, Nova Scotia.*

The saving moment of the passage comes at the end of the day on Tuesday as *Brendan* turns the buoy at Cape Canso and runs down the last twenty miles into Saint Peter's Bay. The fog is thick as night. The wind is still fitful, frustrating our attempts to shut down the engine and sail the final leg. But as loud and monotonous as this long motor sail has become, the crew of *Brendan's Isle* rises beautifully to the occasion.

Each person takes a job, competently and without complaint. Peter prepares a fine supper of chicken stew and salad. Mike and Blue work the foredeck, handling sails, stowing gear, sharing the anchoring duties. Amanda teams up with me: she stands behind the helm wearing a battery-powered walkie-talkie while I sit in front of the radar wearing the other walkie-talkie, conning her past the maze of entrance buoys and into a cove on the south side of Saint Peter's Bay.

The evening is so thick that even the foredeck crew never see the loom of the land. Amanda slow the boat and feels her way toward the beach with the aid of an electronic depth sounder. When the depth is about twenty-five feet, she drops the engine into reverse and calls for Mike and Blue to let the anchor run.

On a signal that the anchor is set, Amanda shuts down the engine and I switch off the radar and climb to the cockpit. An uncanny silence seems to envelop the boat—a counterpoint to the engine noise that we've been forced to endure for nearly twenty-four hours. Slowly, as our ears become accustomed to the silence, other, quieter sounds begin to emerge. A heron screeching somewhere behind the beach. The hissing of surf on the far side of a nearby spit of land. The sound of a distant foghorn.

Three days later and seventy-five kilometers farther north, *Brendan* lies anchored a few meters off the end of the government pier in the harbor of Baddeck. The crew splits up after breakfast to work on last-minute rigging and provisioning chores. I dinghy ashore with the others, then turn and start following the harbor road toward the west end of town.

A quarter mile along this road, I stop for a moment and look back at the sailboat. I gaze at her dark green hull floating in the mirror of its own reflection, and I start thinking about all the distance that she still has to sail this summer. She's obviously rigged for sea, with a tall single mast, tandem jibs, a pair of heavy downwind poles mounted vertically on the foredeck. Her high flush decks, her flared clipper bow, and her round canoe stern give her a utilitarian look, as if she's been built with work in mind. An automatic steering apparatus, called a steering vane, is bolted to her stern, and a radar mast and several radio antennas are arranged in a semicircle around her cockpit.

The boat looks strong—and indeed she is strong, under

most ordinary circumstances. But her hull is made of fiber-glass—not steel—and "most ordinary circumstances" for a hull of this type do not include the possibility of collision with floating slabs of ice.

Last fall as I was fitting her out for this voyage, I thought about whether to reinforce her bows with wood or metal sheathing as a way of strengthening them against an accidental collision with the ice. I talked with a number of others—engineers, surveyors, boatbuilders—before coming to the conclusion that in this case sheathing wasn't going to do much good. As one wise old boatbuilder explained, it would be a little like putting a Band-Aid on an eggshell: the protective layer might guard against superficial nicks and gouges, but the basic structure of the eggshell underneath (in this case, the fiberglass hull) would still have to withstand the impact of the blow.

So the decision was made and *Brendan's Isle* was not ice-strengthened. Does this mean that she's the wrong boat for the job? Some might argue so, but I would disagree. She might be the wrong boat if our intention this summer were to pick a particular destination and just sail there, ignoring the ice, smashing through it if we couldn't find a way around. She might be the wrong boat if we had a series of fixed and immutable deadlines or if we needed to operate on a published schedule like some sort of commercial eco-tour.

But I think she's exactly the right boat for doing what we are sailing north this summer to do: to learn about the ice by making ourselves vulnerable to it—so that its size and thickness, its concentration, its present and future whereabouts become vitally important to us. In a fiberglass boat like *Brendan's Isle,* we will be able to sail *to* the ice but we won't be able to sail *through* it. What this means is that we'll have to let the ice become our teacher—learning to live by its rules rather than our own.

————

By afternoon most of the preparations for the next leg of the voyage have been completed and I'm on my way to the local boatyard office to pick up two fax transmissions: the latest ice analysis from the Navy/NOAA Ice Center in Washington, D.C., and a four-day weather briefing from meteorologist Bob Rice. By the time I return to the boat, Mike has collected the day's Canadian ice maps from under the radio-fax machine, cut them into appropriate sizes, and gathered the crew around the dinette table for the weather and ice briefing that we've all been waiting for.

The weather looks good: settled conditions with forecast moderate westerlies. Pretty much the way it's been up here for the past month. We may not have a lot of fast sailing for a couple of days. But what wind there is will be behind us—and the seas will be gentle.

So the weather forecast says "go." Now the question is . . . "go where?" Southeast around Newfoundland? Or northeast through the Strait of Belle Isle?

I've purposely not looked at any of these maps for several days, hoping in this way that evolving trends might appear more obvious. But it's evident almost at once that no such precaution has been necessary. The evolving trends are clear for everyone to see.

Both the U.S. and Canadian charts show the ice moving farther offshore and continuing to decay. There is no longer any significant sea ice being reported east of Newfoundland—just a few scattered patches in the eastern approaches to the Strait of Belle Isle. The only meaningful concentrations of ice in the entire area are located in wind and current rows, called "ice bands," that stream out to sea for a hundred miles from the mid-Labrador coast.

I admit that I'm surprised. It seems my crew and I are looking at a condition very close to the climatological norm for sea ice in early July—a condition that has not occurred in this part of the Labrador Sea for five years and has occurred only one other time in the past ten years. If the ice here in 1991 was the heaviest in a century, the ice in 1994 looks to be the lightest in a decade.

Based on these reports, there is little question about which route *Brendan* will sail to Greenland. With the small amount of ice cover shown on these maps, she could safely transit the Strait of Belle Isle today. In another week, if the ice continues to decay, she should be able to sail in ice-free waters halfway to Baffin Island before being forced to set out across the Davis Strait.

A fine mist begins to fall just before the dinner hour, causing a hasty conclusion to our deliberations in the main saloon. Blue and Amanda hurry to retrieve laundry hanging in the rigging. Mike hauls down a pair of sleeping bags draped across the main boom. Pete rescues his plankton net, test tubes, and petri dishes from the cockpit, where earlier he'd been observing his most recent sampling of the microbial population of Baddeck harbor.

Sunrise will come early tomorrow—before five o'clock—and the current will turn in our favor soon afterward. We will depart from Baddeck before breakfast for the twenty-mile run down a long, wooded fiord called the Great Bras D'Or. Then we'll pass through the tidal race at Aconi Point and proceed northward into the Gulf of Saint Lawrence.

I feel lucky tonight. The vagaries of wind and current (and who knows what other forces?) have opened a route that may save hundreds of miles and several days of sailing time. If the good weather holds and ice conditions in the Labrador Sea continue to improve, the way will be clear for a dash along the west coast of Newfoundland to the Strait of Belle Isle: our shortcut route to the ice.

# II.

# North Northeast

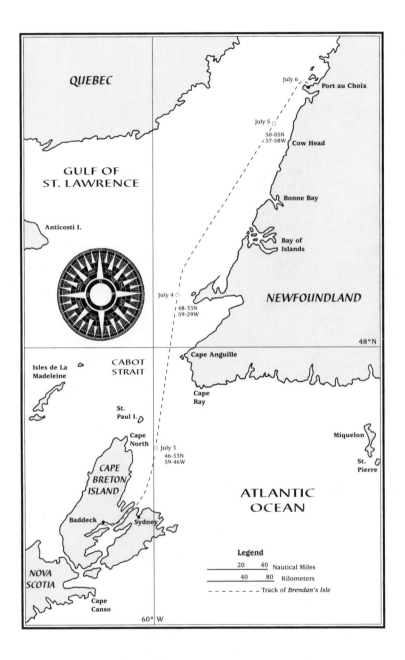

QUEBEC

GULF OF
ST. LAWRENCE

Anticosti I.

NEWFOUNDLAND

July 6

Port au Choix

July 5
50-05N
57-58W

Cow Head

Bonne Bay

Bay of
Islands

July 4
48-33N
59-29W

48°N

Cape Anguille

CABOT
STRAIT

Isles de La
Madeleine

Cape
Ray

St.
Paul I.

Miquelon

Cape
North

July 3
46-53N
59-46W

St.
Pierre

CAPE
BRETON
ISLAND

ATLANTIC
OCEAN

Baddeck

Sydney

NOVA
SCOTIA

Legend

20        40   Nautical Miles

40        80   Kilometers

- - - - - - -  Track of *Brendan's Isle*

Cape
Canso

60° W

*July 3, Sunday afternoon. 46/53N, 59/46W.*

*LOG: Wind: west, 10 knots. Barometer: 30.01 rising. Weather: clear, cold. Position: Gulf of Saint Lawrence, 52 miles SSW of Cape Ray, Newfoundland.*

As we make our way out into the Gulf of Saint Lawrence this afternoon, running before a westerly breeze under a full press of sail, we are a halfway house between the old and the new. As a sailing vessel, we are the inheritors of an ancient tradition, moving across the surface of the sea by means of a technology that is passive and participatory. In spite of some of the fancy new gear on board—the self-tailing winches, for instance, or the roller-furling headsails—we are a vessel whose technology is fundamentally the same as it might have been four hundred years ago—a vessel that John Davis himself could have stepped aboard and sailed without too much difficulty.

Yet at the same time we are the end users of the most modern electronic technology. We navigate primarily by GPS, an instantaneous satellite-generated global positioning system. We make decisions about route and timing based on a stream of satellite-derived weather and ice reports, analyzed by technicians thousands of miles away and delivered to us by radio facsimile twenty-four hours a day. Even as our physical bodies are raising halyards and adjusting sheets and reefing sails, our electronic "eyes" are circling the Earth, feeding us information about where we are and where we're going from a series of vantage points hundreds of miles overhead.

The odd thing about this combination of opposites is that both ends of the technological spectrum are essential to the purposes of this summer's voyage. The technology of sail is silent. It doesn't "use up" anything. It doesn't change anything chemically or leave a trail of waste in its wake. As a metaphor, it thus serves as a statement about how we might learn to coexist with the rest of nature without littering the planet with our carelessness. As a method of travel, it also functions as a passive technology, like a windmill or a water-wheel, whose object is not to overcome the forces of nature (as the engines of the industrial revolution do) but to coop-erate in interactive partnership with them. As such, it is a means of carrying us into the natural world and allowing us to contact it in a particular way—as participants—learning to defer to nature's requirements and to subordinate our own.

In the context of this summer's voyage, however, none of these advantages of traveling in a sailboat would be enough without the additional advantages of the many earth-orbiting eyes that we carry. Strictly speaking, the information that comes from the GPS satellite navigator is not necessary, as nearly everyone on board is trained in the use of a sextant. But all the other electronic information we receive is neces-sary indeed. Not only does it provide a basis for making day-to-day tactical decisions, it also gives us a means of perceiv-ing ourselves in relation to much larger hemispheric patterns and thus of thinking in a more holistic way about the Earth and the forces at work upon it.

———

There are three large hardbound volumes aboard *Brendan's Isle* this summer, all having to do with satellite observations of global systems, all gifts to our expedition from one of their authors, NASA climatologist Claire Parkinson. Together, these volumes represent nearly twenty years of collaboration by a team of NASA scientists, many of whom have spent their professional careers observing and analyzing remote-sensed data about Arctic ice and climate. My thoughts this morning about our dependency on electronic satellite eyes gets me to thinking about these volumes, about Claire Parkinson, and about the important breakthrough that has taken place during the past twenty years as a result of satellite surveillance of the polar regions.

Climatologists have long understood that sea ice plays a key role in the dynamics of Earth climate. Yet until recently they have had no way to observe the ice regime as a whole or to gather information about it as a unifed component of the global climate system. Fifty years ago, most data that existed about polar sea ice still had to be gathered from a few scattered observation stations or from dangerous polar expeditions. A few decades later, the development of technologies such as infrared photography and side-scanning radar enabled scientists to begin collecting a more consistent body of information. But it wasn't until 1972, with the launch of a satellite-mounted passive microwave imager called the Electrically Scanning Microwave Radiometer (or ESMR), that the real breakthrough came. The launch of the ESMR meant that for the first time in history scientists could observe the polar ice without interruption from weather or polar darkness and could begin to analyze its large-scale structure, motion, and variability.

The importance of satellite monitoring of the polar sea ice

is now well understood. And everywhere there is mention of such monitoring, the name Claire Parkinson is somewhere close at hand. As a research scientist since 1978 at NASA's Goddard Space Flight Center near Washington, D.C., Parkinson has been actively involved in project after project relating to passive microwave imaging of both Arctic and Antarctic sea ice. She has served as lead author of an atlas of the ESMR data, and she has written or collaborated on dozens of other books and scientific papers about her work at Goddard and elsewhere.

Over the past several years Claire Parkinson has become an important influence in my own evolving interest in the study of sea ice and Earth climate. Soon after I started exploring the scientific literature, I began noticing her name cited in dozens of key publications. Then I read a pair of her articles, both relating to sea ice and climate change, and I realized that I needed to meet this person face to face and talk with her about her work.

Since that time I've made numerous visits to Goddard. On each occasion Parkinson has been a patient listener, an asker of pointed questions, a critic of sloppy thinking, and a wellspring of information about the most important issues and the most influential researchers working in her field. Over the years she has become both a mentor and a friend. Without her help it is unlikely that either this investigation or this voyage would ever be happening.

One of the most important dimensions of Parkinson's work—along with that of her colleagues at NASA—has to do with the way it has changed our thinking about the planet we live on. A few decades ago researchers were still trying to understand the Earth by looking at it through a series of separate windows. Then along came two revolutionary new tools—the Earth-orbiting satellite and the computer—and a new generation of scientists trained in their use. The satellite was pressed into service as an Earth observatory, while the computer found one of its most important applications as a modeling device, capable, in the hands of a trained mathematician, of simulating the behavior of very large, complex, nonlinear systems such as the Earth's climate. Together these tools enabled researchers to begin thinking about the Earth in a new way. A new "cybernetic" consciousness was born—and along with it a new understanding of the planet we live on as a single, highly complex, interactive system.

Like NASA's internationally known global climate modeler Jim Hansen, Claire Parkinson is a talented mathematician, trained in computer modeling of Earth climate. But unlike Hansen, Parkinson is not a predictor of futuristic scenarios. In fact, in spite of her vantage point at the forefront of climate study, she is generally unwilling to speculate about future events based on her (or anyone else's) research.

This conservative cast of mind derives from her background in mathematics and her basic desire to prove things irrefutably before making any claims. But it is also grounded in dozens of scientific examples. During our conversations she gave a number of such examples relating specifically to the field of climate studies. The one that struck me most strongly, however, came from her own experience.


Her story begins in 1980, two years after she first arrived at Goddard. This was the period when she and others had just started investigating the ESMR record for patterns, relationships, anomalies, anything that might stand out as scientifically noteworthy. Working with a Princeton undergraduate named Andrew Gratz, she soon discovered a series of curious events that had occurred in a small, nearly landlocked sea near northeastern Siberia—the Sea of Okhotsk.

The satellite record for this region showed the sea surface with more than ninety percent ice cover during February 1973, whereas the record for the following three Februarys showed up to forty percent less ice cover in the same area, with much of the central and eastern portions completely ice-free. The potential significance of these findings increased dramatically as Parkinson and Gratz came across a second set of data from the National Center for Atmospheric Research that correlated almost perfectly with the sea ice record. A review of these data revealed that during all three low-ice years, the air flow in the Okhotsk region had been dominated by prevailing winds from the relatively warm Bering Sea area. In contrast, during the single high-ice year, the air flow across the region had been dominated by winds from the much colder area of northern Siberia.

In response to these findings many scientific observers might have been tempted to hypothesize a causal relationship between wind and ice. The correspondences appeared strong, and their explanation seemed highly plausible. Consistent with her conservative tendencies, however, Parkinson avoided predicting future behaviors of any sort.

"I suppose we could have generalized these findings," Parkinson said, "but there was no way that I would have been comfortable saying anything beyond the fact that the correspondence between the two data sets looked suggestive. Too many things could change as we started looking at additional data."

Parkinson's Sea of Okhotsk paper was published in 1983, by which time Parkinson and other NASA scientists had already started looking at another set of data from a new microwave imager called the Scanning Multichannel Microwave Radiometer (or SMMR). A few years later, after almost nine years of SMMR data were available, she turned her attention once again to the Sea of Okhotsk, repeating the steps that she and Gratz had followed for the ESMR data, making detailed comparisons between the sea ice cover and prevailing winds.

By 1990 the verdict was in, and—perhaps not so surprisingly—the restraint she and Gratz had exercised in their earlier reporting was borne out by the new results. As strong as the wind/ice correlation had appeared during the ESMR years, the new data showed almost no such correlation. The evidence had been circumstantial and the relationships between ice cover and air temperatures merely coincidental. Parkinson's cautious treatment had turned out to be not only the most conservative but also the most scientifically responsible course of action.

For Claire Parkinson, the Sea of Okhotsk story serves as a dramatic reminder of just what the passive microwave sea ice data are—as well as what they are not. They are exciting, promising, new. They are science's first large-scale glimpse at a geographically diverse component of the Earth's climate system. But they are also limited, both in duration and in the type of information they provide.

"It's important to keep in mind that the ESMR data spanned a period of only four years," Parkinson says, "—also that the ESMR was a single-channel instrument without the ability to distinguish between various types of ice. The SMMR record is longer—almost nine years instead of four—and it contains much more information from several different channels. Still, when measured against climatological time scales, the SMMR records too are statistically very incomplete."

The point, for Parkinson, is that it's going to require a much larger, statistically meaningful sample before researchers will be able to distinguish between simple coincidence and consistent, repeatable patterns in the year-to-year ice record. Concerned observers all over the world are asking to know *now* about what's going on in the polar ice and what it may mean, and Parkinson understands the urgency of such requests. But as she never tires of pointing out, science has its own timetable. If we're looking for correct answers that will stand up to long-term scrutiny, she warns, we may simply have to wait until more of the evidence is in.

*July 4, Monday afternoon. 48/33N, 59/29W.*

*LOG: Wind: southwest, 8 knots. Barometer: 29.90 steady. Weather: increasing high cirrus. Position: 10 miles W of Red Island, Cape Cormorant, Newfoundland.*

The land we see this afternoon rises a thousand feet out of the sea and stretches like a long gray battlement along the eastern horizon. A steep-sided island of reddish sandstone stands like a sentinel three miles from this coast and serves as the only distinguishable landmark to measure *Brendan*'s progress along the shore.

The afternoon is clear. The boat moves easily, with wind and seas on her quarter. By shortly after one o'clock the air temperature has warmed to eighteen degrees Celsius (almost sixty-five Fahrenheit), and *Brendan*'s crew is scattered about the decks, reading, resting, trying to take advantage of the warmest temperatures we may experience at sea for the rest of the summer.

I sit by myself on the foredeck staring vaguely ahead, still pondering the stories of Claire Parkinson and the two NASA sea ice studies that have filled my thoughts since yesterday afternoon. Parkinson's candidness about the limitations of these studies points inevitably to a number of questions—questions that nagged at me until I finally found the opportunity to pose them to her during one of our meetings at Goddard.

If the sea ice data are indeed so short, I asked her, then what is their scientific value? How are researchers supposed to use them? If four years isn't enough time for the data to confirm a pattern, is nine years enough time? Is thirteen years enough time? Who decides how much time is enough time?

And what about the geographical region we've been talking about—the place I've been calling the LDB area? What do the NASA data sets have to say about this region? What do they have to say about the Arctic as a whole? Do they confirm any of the behaviors that the Canadian ice forecasters and others have observed—the out-of-phase behavior of the LDB area, for instance? Do they *explain* any of these behaviors?

As quick as Parkinson had been to admit to the long-term inadequacies of the two data sets, she was just as quick to emphasize their short-term usefulness. "Thirteen years may not amount to much on a climatological time scale, but with the data from these thirteen years, we've accomplished a lot," she said. "We've compiled records for both hemispheres on a daily, monthly, and yearly basis. We've established baseline values, calculated trends, made comparisons with ground and aircraft observations. In spite of the limitations of these data as predictive tools, we've been able to assemble them as benchmarks—so at least we have a place to begin as we try to understand what's been happening in the ice."

In answer to my questions about the LDB area, Parkinson described a project that she and a colleague named Donald Cavalieri had collaborated on during the late 1980s: an analysis of the combined ESMR/SMMR data for the entire

Northern Hemisphere. This analysis, although necessarily limited in time, contained a computation of yearly averaged sea ice extents for eight subregions of the Arctic as well as for the Arctic as a whole, and it helped identify a number of regional ice behaviors that had not been well understood before.

"Our analysis of the two data sets provided a mixed result," said Parkinson. "During the four years of ESMR data, we found considerable variation in the eight subregions, with a total indicating a slight increase in ice cover across the Arctic as a whole.

"But when we looked at the SMMR data, we found a decrease in overall ice cover. Two of the subregions, including Baffin Bay/Davis Strait, had clearly increasing trends, three had changed very little, and the other two had clearly decreasing trends."

A question that I asked Parkinson almost immediately was whether she thought the overall decrease in Arctic ice cover during the SMMR years might be large enough to be considered a climate signal. Although the change was not as large as those forecast by many of the General Circulation Models, it was nonetheless in keeping with several GCM predictions. It was also consistent with a number of other changes currently taking place around the Earth and identified by the GCMs as possible signals of climate warming: a marked increase in global atmospheric temperatures over the past three decades combined with a twenty-year record of decreasing stratospheric temperatures; a warming of the Alaskan permafrost during the SMMR lifetime of 0.5 degrees Celsius; a global sea level rise of one or two millimeters per year over the last century and three to four millimeters per year since 1990.

Parkinson agreed that the decreasing sea ice cover was in keeping with the predictions of many of the GCMs. Once

again, however, she advised caution in interpreting this finding. "Even though the direction of change is downward, the
*rate* of change is still small. Without a stronger indicator from
the ice, it may be impossible to differentiate between this
small rate of change and what climatologists call 'background
noise'—the ebb and flow of normal climate variability."

"But what about the out-of-phase behavior that you and
others have observed in the Baffin/Davis region?" I asked.
"What if increases in the ice cover here are actually some
kind of upside-down response to regional warming—as
some of the Canadian forecasters have been suggesting?
Then shouldn't the sign be reversed so that this change is
*subtracted* from the Arctic total instead of added back in? And
wouldn't this make the warming signal coming out of the
Arctic a lot more dramatic?"

"It would certainly make the signal more dramatic—but
I'm not sure it would make much sense," said Parkinson.
"The answer to your question depends partly on the mechanism that's causing the out-of-phase behavior. Unfortunately, the satellite data don't provide answers about mechanisms. They're excellent for observing large-scale variations
in the sea ice. But they don't explain *why* those variations are
taking place."

"All right," I said. "I'll accept that. It's like the Okhotsk
problem—the satellite data by themselves may not contain
enough information to explain the observed behaviors. But
it seems that here the situation is a little different. Because
here we're not talking about just one anomalous year—we're
talking about an entire anomalous region—a *consistently*
anomalous region that may have been out of phase for decades, maybe even for most of this century."

"So?"

"So it seems that until scientists are able to identify the
mechanism that's causing this out-of-phase behavior, they

won't know whether to add or subtract it from the total. Which means they won't be able to place a value on the changes taking place in the Arctic as a whole—or to understand what they mean."

Parkinson gazed at me for several seconds before she responded. "If you want to answer questions about the physical forces at work in the ice, you'll need to talk with scientists whose work is specifically focused in this area: oceanographers, geochemists, geophysicists. Look at their published work. Then go meet with some of them. Find out about the questions they've been asking and the answers they've proposed. See if any of them has found a plausible explanation for the Baffin/Davis anomaly."

I thanked Parkinson for her advice. She was right, of course. She was only the first of many researchers I would need to contact as this investigation unfolded. The NASA sea ice studies were important—but they contained only one kind of data and shed light on only one part of the riddle. Next, I decided, I would need to seek out the person whose ideas I had first heard from an anonymous Canadian ice forecaster during this summer of 1991—the researcher who had identified a set of sea ice behaviors in the Baffin Bay side channels that might be responsible for the out-of-phase character of the entire LDB area.

*July 5, Tuesday. 50/05N, 57/58W.*

*LOG: Wind: southwest, 28 knots. Barometer: 29.08 falling.*
*Weather: thickening low stratus cloud, dropping visibility. Position:*
*14 miles WSW of Portland Hill, Newfoundland.*

All during the afternoon on Monday the sky grew darker, and by evening the wind had backed and had started to increase. As Tuesday's dawn breaks, I find myself staring at a stack of weather fax reports that have come in during the night, wondering what will happen next. According to these reports, a low-pressure system that has been nearly stationary for the past several days has started to deepen and head our way. Gale warnings are posted for the Strait of Belle Isle in advance of this system tonight—storm warnings are possible for tomorrow.

After careful consideration, I decide to carry on. The wind and seas will undoubtedly mount during the day, but *Brendan* behaves well in a following sea, and we've got twelve hours until the real weather is due to arrive. Seventy-five kilometers ahead on the northeast peninsula of Newfoundland is the village of Port au Choix, a well-protected harbor that is the base of operations for the local fishing fleet. Once this weather passes, Port au Choix will also serve as a useful staging point for transiting the strait.

On deck the wind rises. Amanda and Pete move forward to tuck a reef in the mainsail—a maneuver they've not yet performed with the boat surging and rolling hard. They work slowly, talking together through each step of the process. Blue, meanwhile, stands at the helm, hand-steering. He purses his lips and stares straight ahead, concentrating on keeping the rudder centered as the boat skids downhill.

Ten miles to the east, Newfoundland has become a hulking giant of dark green mountains and steep-sided gorges that slice deeply into the interior. The mountains rise in silhouette against the sky: Gros Morne, Western Brook Peak, Indian Lookout, Gros Pate. They are the highest mountains in Newfoundland, the backbone of the western side of the island and the northernmost evidence of the Alleghenian orogeny, a geologic episode that formed the Allegheny and Appalachian Mountains, extending all the way back to Georgia and defining one edge of the North American craton.

The gorges are actually drop-faulted valleys, scoured and enlarged by aeons of glacial scarring, footed in freshwater lakes, and separated from the waters of the gulf by a coastal plain of till and terminal moraine. They open one by one as we pass—Western Brook Pond, Saint Paul's inlet, Parson's Pond—then close again like secrets the mountains do not want to share.

The sky drops lower. The mountaintops lose their definition. Scuddy wind clouds stream above the coastal plain, defining a shear zone where the wind increases in a venturi effect, blowing a fine gray mist along the shore and whipping up the surface of the water. The boat accelerates in the north-flowing current, making better than nine knots over the bottom.

Five hours, I think. Five hours and we'll be around the corner at Pointe Riche lighthouse and running down the lee side of the Port au Choix peninsula—plenty of time to get into the harbor ahead of the gales that are forecast for to-night. Meanwhile, *Brendan* surges faster and faster on mounting seas, oscillating through deep power rolls that sound like small explosions down in the cabins, while her crew settles in for a wild sleigh ride down the final miles of the Gulf of Saint Lawrence.

*July 6, Wednesday. Port au Choix, Newfoundland.*

Six summers ago when *Brendan's Isle* first sailed into Port au Choix, this place was still a boom town. The largest and most prominent building on the waterfront, the FPI fish-packing plant, was operating at full tilt—two full shifts a day, six days a week. Forty codfish draggers, most with local skippers and crews, were operating out of the area and selling to the plant. Five refrigerated tractor-trailers a day were carrying frozen codfish fillets down to the ferry terminal at Port aux Basques. Dragger captains were building two-story houses with their profits—houses that looked like the ones rich American suburbanites lived in, with paved driveways, basketball hoops, and electric baseboard heaters.

That was the summer of 1988. By the fall of 1991 the federal government in Ottawa had finally admitted that there had been a "catastrophic decline" in codfish populations all around Newfoundland, and six months later it made the decision to close the fishery until further notice. Almost overnight, villages such as Port au Choix became economic disaster areas.

Last evening when we arrived in the harbor, we moored *Brendan* at the eastern end of the fishermen's cooperative pier, inboard of about twenty local codfish draggers, most of them mothballed and unattended. Across the harbor on the far side of the village was the FPI fish plant. From a distance the plant looked the same as it always had—a tidy, utilitarian two-story rectangle sheathed in gray corrugated metal. But from closer up the changes were apparent. The paint on the walls and roof had started to peel. Several of the doors looked as if they'd been forced in. There was trash piled in the parking lot and broken wooden pallets in the shipping bays. The problem was obvious: nobody was working in the plant this summer. The fish that used to be processed here were gone and the plant had been closed—all but a few weeks a year—since the codfishing moratorium had been imposed in the spring of 1992.

By the time we were finished setting up bumpers and adjusting mooring lines last night, the wind was blowing hard, making a low, whining noise in the rigging and driving a fine mist so hard it hurt your face when you looked into it. There were no other people anywhere along the waterfront. The unpaved streets of the village were deserted, too, and as the rain increased, the streets began to run in muddy rivulets and to fill up with large brown puddles.

Now, twenty-four hours later, there are breaks in the overcast, the rain has stopped, and a few people have started to move around the pier. They are fishermen, mostly, from a couple of the working draggers moored ahead of us. One is an older man—in his late sixties or early seventies—with matted gray hair, white sideburns, and a white stubble beard. He stands next to a large steel dragger called *Three Brothers*, and he sniffs the air and looks out to the west, watching the clearing come.

I'm also walking on the pier this afternoon, doing just

what this man is doing—stretching my legs after a night of being stuck in a little cabin. I stop a few paces from him and gaze up at the high flared bows and the rust-streaked steel superstructure of the dragger in front of us.

"Nice boat," I say.

"Aye," he says.

"Are you her skipper?"

"Aye, m'son."

"What're you fishing? I thought codfish was illegal now."

"Illegal or no, I s'pose it don't matter," he says in his Newfie brogue. "There's no more fish. And no place to sell. We gets a little redfish now and again. And shrimps, too, as soon as the season begins. But 'codfish'—as you calls 'em—is scarce, very scarce."

I want to ask this man the hard questions: Why he thinks the codfish are gone. Whether he thinks they'll ever come back. What's going to happen to the people in villages like this one over the next few years. But I know better than to approach these subjects head-on. Instead, I introduce myself as the skipper of the green sailboat moored down at the end of the pier.

He asks where we're from. I tell him we're from the States. He asks why we're here. I say we've come to look into some of the things scientists are saying about the sea ice in this area and about global climate change.

These remarks seem to open up a verbal floodgate. The man, who calls himself Kevin Simms, begins with the demise of the cod fishery several years back. It was terrible, he says. Worse than anyone who didn't live here could possibly imagine.

"There's many here that blames it on the seals—or the Russians or the Japanese. But to my way of thinking, we was to blame for it—for the greater part of it, anyways. The problem, if you wants to know—the problem that started it all—is right here." (He gestures with a sweep of his arm toward the row of draggers moored in front of us, his own included.) "The boats is too big. The trawls is too wide. The mesh on the nets is too fine. There's too much fish gets dumped back in the sea at the end of a haul.

"And now there's something else—something funny going on. Because now two years after, the fish still ain't coming back. They just ain't growing up the way they once growed. A five-year-old fish ought to be twenty inches cheek to tail and weigh four, maybe five pounds. But now, the scattered one we finds in our nets is all skin and bone, no meat at all. When you holds 'em up to the light, all you sees is their ribs. Fact is, a man today barely gets a meal from a fish that used to feed a family.

"The bait food is gone, is what I thinks. And I'll tell you why I thinks it. The winter is become too cold of late. The ice stays in the bays too long. Birds that used to come and nest in June—they don't come now until July. The capelin that once run up the bays in May and June now run in July and even August. The winters come too early. And there ain't near as much snow, but much more cold and frost.

"Twenty year ago a man and his crew started fishing in these parts by April, maybe early May. Now, even in a good year, we don't get out until June. And the ice is still too heavy, water still too cold.

"Something's happening, that's sure. The scientists ain't telling us what—maybe they don't know. But I knows. We that lives here knows. I've fished this gulf every summer for fifty-five year. And I knows. Something's wrong. Something funny's going on and they ain't telling us what it is."


Behind the old man the clearing grows. The light that had started as a sliver of orange now expands into a blaze of yellow and purple and gold in the western sky. Patchy breaks in the overcast combine overhead until the blue stretches back to the horizon. A dry west wind begins to raise a chop on the surface of the harbor, and the waves slap against the hulls of the fishing boats.

Kevin Simms looks up at the sky and cocks his ear to the rising wind. I look up as well—then back at *Brendan*. I see Blue and Mike working on the deck, and I know I must be getting back. "Maybe things will change," I say. "Maybe the fish will recover if they're given enough time."

"They was so many . . ." the old man says, speaking more to himself than to me. "We never thought they'd ever come a day . . . we just never imagined . . ."

His voice trails off and he turns away. Then he looks down the pier, ignoring me, and he calls to one of the men who's been standing near the brick cooperative building a few dozen meters away. "Let's get them nets aboard now, Billy, and them booms rigged down. It's nearly time to go."

With the clearing coming, I head back to *Brendan* to check the forecast and to see about gathering the crew. If the weather looks right, I know we'll also be sailing soon.

On the boat, Mike is already several moves ahead of me. He's sent Pete off to the village to pick up a few last-minute supplies and to look for Amanda. Meanwhile, he and Blue have cleaned the cabins, reorganized the sail locker, set a pot of chicken on the stove for supper. He's also copied out the local forecast on a sheet of paper and set it on the navigation table for me to read.

The forecast speaks even louder than the sky: moderate northwesterlies tonight, going light by early tomorrow. There's no need for discussion: as soon as the others are back on board, we sail for the strait.

# III.

# Strait of Belle Isle

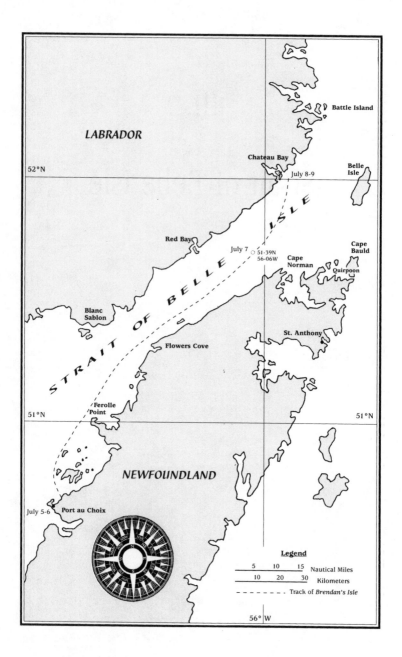

LABRADOR

Battle Island

Chateau Bay

52°N

July 8-9

Belle
Isle

Red Bay

S T R A I T    O F    B E L L E    I S L E

July 7    ○ 51-39N
                56-06W

Cape
Norman

Cape
Bauld

Quirpoon

Blanc
Sablon

Flowers Cove

St. Anthony

51°N

Ferolle
Point

51°N

NEWFOUNDLAND

July 5-6    Port au Choix

56° W

*July 7, Thursday afternoon. 51/39N, 56/06W.*

*LOG: Wind: nil. Barometer: 29.88 steady. Weather: broken overcast, air temperature 7C, water temperature 5C. Position: Strait of Belle Isle.*

Until you've crossed the boundary of a major oceanic current, it may be impossible to appreciate the abrupt change in climate that such a boundary represents. The most dramatic example of this kind of change in the temperate latitudes is the Gulf Stream. In the tropics, where the Gulf Stream originates, there is no apparent boundary. But north of Bermuda the boundary hits you like a wall, with an abrupt change of water and air temperature, confused seas, towering cumulus clouds, strong winds, thunder, lightning flashing overhead.

The boundary of the Labrador current, although generally less visible in the atmosphere, is no less dramatic. Fifty kilometers to the west, the shores of the Gulf of Saint Lawrence are covered with boreal forests and clad in summertime green, with winter snows and coastal fast ice long since melted and snow patches relegated to the highest mountain peaks. Then, within a span of thirty kilometers as you transit the Strait of Belle Isle, the water and air temperatures plummet, the tree line recedes inland, and the coastal bays and headlands take on a harsh, Arctic appearance, with outcroppings of ancient Pre-Cambrian rock scattered over the Labrador hills and bogs of peat and Arctic tundra lining the Newfoundland shore. Snow patches appear everywhere along the beaches and in the crevasses of hills. And in the eastern approaches to the strait, where the Labrador current sweeps around in a great lazy eddy, dozens of icebergs dot the horizon, trapped in a cul-de-sac where they wander in circles among complex tidal currents.

Just before noon, *Brendan* completes her transit of the central narrows and enters this area of bergy water. The number of icebergs visible around the horizon has grown with every kilometer—first three, then six, then ten, then twenty and more. An odd silence has come over the crew. The boat is under power, moving directly toward the largest of the bergs—a twin-towered monster that Mike and Blue first spotted earlier this morning. Amanda is at the helm. At five hundred meters from the berg she throttles back the engine. At three hundred meters she throttles back again and looks at me, waiting for a signal to proceed.

Funny, I think—she and I are the only people on board this summer who've ever been this close to an iceberg. Which also means that she and I are the only ones who understand just how deceiving these massive objects can be. They appear inert and profoundly stable. This one measures thirty to forty meters tall and several hundred meters long. It is fronted on one end by a massive ice cliff and on the other by a beach where the swell creates a long, curling surf. Gulls nest on the nearest peak; a pair of seals bask on the ice beach; terns fish in the shallows where a fall of meltwater pours from a glittering promontory. Except for its color, the berg looks like an actual island, as fixed and permanent as any land.

The difference, of course, is that this land is floating, balanced on a point of temporary equilibrium in which the center of the mass is somewhere far below the water surface, low enough to maintain a stable attitude. The second difference is that this land is also *melting*—which is to say that it is constantly changing its point of (temporary) equilibrium.

Amanda knows what this can mean. Several years ago when she last sailed to Labrador, she witnessed, with me, an iceberg of similar size literally destroy itself in a matter of minutes. Beginning with a single explosion, one section of the berg we were watching suddenly broke away and slid off into the sea. The entire mass then began to roll, trying to find a new point of equilibrium. As one visible portion began to sink, another that had been underwater began to rise several hundred meters away. Then, as can sometimes happen when a berg is rotten and full of shock faults, another section broke away, and another, and another. In a paroxysm of explosive calvings that set up a series of tidal waves and echoed like cannonades against the shore, an iceberg that only ten minutes before had been the size of a small office building proceeded to blow itself to pieces.

I look at Amanda, wondering if she is remembering the same event. We won't approach too close, I promise her— just close enough for Blue to get a few photographs and for Pete to hang from the boarding ladder and try to retrieve a floating piece of ice for the galley icebox. When Blue begins to complain that we're not close enough, I wink at Amanda. Then I suggest that before he gets feeling too brave about approaching this immovable island, he might ask Amanda to tell him her favorite iceberg story.


Once through the narrowest section of the strait, we have several options for this evening's anchorage. A few miles due north is Red Bay—site of the sixteenth-century shore-whaling station that the Basques once called Buterus. Fifteen miles to the southeast is Cape Bauld, Newfoundland, and a good, all-weather harbor called Quirpoon. And twelve miles to the northeast is a series of islands and protected coves called Chateau Bay, the last anchorage but one that *Brendan's Isle* visited in 1991 before she was forced to turn around.

Perhaps for sentimental reasons, I decide on Chateau Bay. I want to return to the place where our voyage was thwarted three years ago. I want to climb the basaltic cap on Castle Island and look over the strait again toward Cape Bauld and Belle Isle and Battle Harbor. I want to gaze out at the ice as it is this year and remember how it was then.

At my request, Amanda points the boat northeast and steers for a series of islands jutting into the strait, marking the mouth of Chateau Bay. The afternoon is sunny and surprisingly warm, and we travel in silence, detouring several times to avoid icebergs that lie haphazardly along the route. I stare ahead at the square shape of Castle Island growing on the horizon, and I find myself pondering some of the questions I'd asked there three years ago about mechanisms in the ice that might be causing the strange, upside-down behavior of this area.

Here we are again in a place whose changing climate doesn't seem to fit any of the current theories about what should be happening. The mean air temperatures here are falling when they should be rising. The sea ice cover is growing when it should be retreating. The whole place is backwards, and nobody seems able to explain why.

This situation gets me to thinking about a similar problem that once confronted the great American geophysicist Maurice Ewing, the man who explored the mid-Atlantic rift and ridge and who eventually discovered the mechanism for the whole plate tectonic theory of continental drift.

Ewing's dilemma involved an odd set of findings that he uncovered during a series of voyages back in the late 1940s when he was doing summer research on the *Atlantis* out of Woods Hole. He was sailing in those years to the central Atlantic to do core sampling of the bottom sediments in the deep ocean. The idea was to verify generally accepted theories about the great geologic age of the oceans. According to what seemed the most logical expectations, the sediments found nearest the continents were supposed to be relatively young, while the sediments found in the central abyss were supposed to be among the oldest on the planet.

The problem was that the actual samples failed to confirm these expectations. The sediments taken from the central ocean near the mid-Atlantic ridge were relatively thin and geologically recent, while the sediments from the continental edges were thicker and increasingly ancient—exactly opposite to what most theories predicted.

Although the idea of continental drift had been around for some time, it was out of favor among scientists because no convincing mechanism for the actual movement of the continents had yet been identified. Thus, all Ewing had to go on were the classic diluvial theories of an ancient ocean. And this meant that the data he was looking at were all backwards, all upside down. The center of the ocean was geologically recent—and nobody could figure out why.

Ewing had the scientific good sense to realize that he was looking at an important finding. He called it a "brutal fact"—the kind of fact that makes no sense but just continues to stare you in the face and refuses to go away. At the time he didn't know what to do with it. What he *did* know was that if he could ever figure out the mechanism—if he could ever determine *why* the ages of the sediments were backwards—this brutal fact was likely to point far beyond itself and reveal powerful secrets about the Earth. (The secrets it revealed turned out to be the whole geophysical process of plate tectonics.)

Could the LDB anomaly be the same kind of brutal fact? I wonder. If science is one day able to identify the mechanism that is causing the out-of-phase behavior of this place, could the answer also point beyond itself? Could the growing ice cover here be a signal, like Ewing's sediments, of something very powerful taking place in the Earth that nobody yet understands?

*July 8, Friday morning. Chateau Bay, Labrador.*

*Brendan* lies at anchor this morning in a landlocked cove. To the east are a pair of islands crowned with square, black outcroppings of columnar basalt—the unique "castle" shapes that give this place its name, both in English and in French. To the north and west are a series of smaller islands that contain a scattering of wood frame buildings and fishing stages—the remains of a village that twenty-five years before had been the home of several hundred people. The buildings are mostly deserted now, and the wooden bridges and footpaths that once connected them are broken and rotted and overgrown with weeds.

Three hundred years ago this place had been a summer fishing station for the large fleets of European codfishing barques and sloops that plied this coast. Fifty years ago it had been a staging point for hundreds of schooners from New England and Nova Scotia and Newfoundland that traveled along the Labrador coast each summer in search of cod and salmon and Arctic char. Now, with all these fisheries either

closed or commercially depleted, the harbor is empty and
the village is a ghost town. Two houses, maybe three, are
inhabited by the sons and daughters of former villagers who
have come back to camp for the summer. Otherwise, the
town is a collection of empty square boxes, bleached and
windswept, with broken windows, doors rattling in the
wind, and graveyards of rotting dories lying upside down in
the yards.

Three years ago, standing on a cliff at the entrance to this
harbor, I'd posed a question about mechanisms in the ice
that might be causing the out-of-phase behavior of this area.
At the time I had only the telephone voice of a forecaster
from the Ice Centre in Ottawa to suggest a possible answer:
an idea the forecaster had referred to as the "side channel
export hypothesis." Then the following winter, after several
talks with Claire Parkinson, I had started searching for someone in the scientific community who might shed a little
more light on this idea—a search that eventually ended at the
doorstep of a Canadian climatologist named John Marko.

Marko and his colleagues, I learned, had been modeling
and analyzing the behavior of the east Canadian sea ice since
the early 1980s and were among the most active researchers
currently working in this geographical region. They were
responsible for developing one of the latest working models
of sea ice advection and iceberg movement around the area.
And what may be most important, they were also the authors of a compelling and little-known hypothesis about the
migration of multiyear sea ice southward through the Baffin
side channels and about its possible effects on ice extents
throughout the LDB area—the so-called side channel export
hypothesis.

The story begins at the office of the Canadian Atmospheric
and Environmental Services (AES) Ice Centre in Ottawa in

the winter of 1991–92. Soon after I had talked with Claire Parkinson that fall, I had contacted this office, looking for more information about the unusual situation in northern Baffin Bay that its forecaster had told me about the previous summer. I needed the name of the fellow I'd talked to. I needed to hear the story of the side channels one more time, then to see some backup material, some data, some scientific studies.

Everybody at the Ice Centre seemed eager to help. Forecasters Phil Cote, Bob Tessier, Claude Dicaire—even the chief of the Ice Centre Dave Mudrey—all were pleasant and cooperative. Nobody was sure whom I'd talked to last summer, but it didn't really matter. What did I want to know? Surely somebody in the office could help.

Unfortunately, good intentions seemed the only help available. Every time I tried to get specific about the subject of last summer's heavy ice and the various mechanisms that might have contributed to it, the dialogue became vague and answers became elusive. Ice Centre staff, it seemed, might be willing to comment about such things off the record, but they were government employees after all, hired to be forecasters, not climatologists. Their job was to deliver information to people like me about *how* the ice behaved—not to speculate about why.

"Okay, fine," I said. "But if you won't talk to me about this subject, who *will*? Who knows about this stuff? Who's studied it? Where did the idea about side channel export come from in the first place?"

Phil Cote eventually provided an answer. "There's a climatologist who's visited this office several times during the past couple of years," he said, "a fellow doing research on the whole east Canadian ice regime. He was here looking at some of the data we have stored at this facility—meteorological records, satellite studies, 'fly-by' data from low-level aircraft reconnaissance. One thing he was particularly interested in was infrared satellite imagery of what we call 'fast ice bridging' in the Baffin Bay side channels. He could probably tell you all about the ice movement up there. His name is John Marko."

Within a week I'd made a telephone call to an independent scientific consulting firm, Arctic Sciences Limited (ASL), in Sidney, British Columbia, and had spoken with Marko. A week later a package arrived in the mail containing a lengthy unpublished manuscript, "Implications of Global Warming for Canadian East Coast Sea Ice and Iceberg Regimes over the Next 50 to 100 Years" by J. R. Marko and others, completed only a few months before and circulated by means of a private printing for the Canadian Climate Centre.

This was the first of many ice studies Marko would share with me over the next three years and the first of many conversations we would have. The study itself was also the strongest statement he and his colleagues at ASL had made to date about the anomalous behavior of the LDB area and about the potential importance of side channel export of Arctic sea ice into Baffin Bay.

Marko, I soon learned, had come to his interest in the ice behavior of this region by a somewhat circuitous route. After receiving his Ph.D. and most of his professional education in the United States, he had moved to Canada to pursue a college teaching career. As often happens to a scientist working in academia, however, he eventually found himself wanting to do science again in a hands-on setting. His background in solid-state physics led to an interest in sea ice dynamics, and this in turn led to a series of research and field study opportunities funded by the oil industry, primarily by Petro Canada.

The first area of interest for these studies was the Beaufort Sea, to the north of northwestern Canada. Then, in 1977,

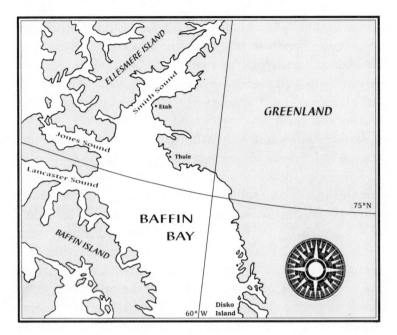

Petro Canada became interested in the exploration of several areas in eastern Canada, in particular a body of water to the north of Baffin Island called Lancaster Sound, the largest and southernmost of the three Baffin side channels.

During the late 1970s Marko received grants for two years of data collection in the Lancaster Sound area, then for two years of analysis and interpretation of those data. It was during this period that he got his first in-depth look at the dynamics and variability of sea ice in Baffin Bay and began to appreciate the importance of sea ice contributions flowing into this area from the three side channels.

Soon after these studies were completed, Lancaster Sound fell out of favor with the oil industry as a site for exploration, in part because of the large volumes of moving ice, in part because of the extreme depths of water (over nine hundred meters) in the channels. It wasn't long, however, before interest was rekindled in another area of eastern Canada, the so-called Hibernia section of the eastern Grand Banks, near the Atlantic coast of Newfoundland. Once again the prospective area of exploration was located dead center in the track of the east Canadian ice regime—and once again Marko and company were called in to prepare a series of feasibility studies and, eventually, forecasting models for both sea ice and iceberg movement through the area.

The early studies relating to the Hibernia project were done primarily to enable oil drilling interests to predict ice volumes that might be moving down the Baffin/Labrador/Newfoundland corridor season to season and year to year. Then, in 1990, Marko and ASL were asked to reevaluate the area with regard to various potential scenarios for global climate warming. (Sadly, the concerns of the oil drilling interests were not the long-term environmental costs of producing ever greater volumes of hydrocarbons to be added to the Earth's atmosphere but the short-term economic costs of losing drill rigs, platforms, pipelines and other expensive equipment to catastrophic, warming-induced ice events.)

The conventional wisdom—as well as the fears of the oil companies—had evolved out of a kind of commonsense scenario of what would happen to the ice as the atmosphere warmed. According to such a scenario, sea ice formation was expected to decrease as air temperatures rose, leading to a reduction in both sea ice extents and the length of the sea ice season across the LDB area. At the same time, iceberg production was expected to increase as warmer temperatures led to greater snowfall amounts and therefore to faster glacial growth.

In their 1991 study, however, Marko and his colleagues took exception to this standard scenario. "The expected high correlations between atmospheric warming and reduced sea ice extents in the LDB area simply don't hold up to scrutiny," Marko explained to me in one of our early conversations. "Such correlations may seem to make common sense. But they are contradicted by the records we now possess of an opposite, out-of-phase behavior that has characterized this area for many decades, possibly all the way back to the early part of this century."

The mechanism that Marko proposed to explain this out-of-phase behavior is based on a phenomenon he first started to notice in the late 1970s during his field studies of the Lancaster Sound area: a phenomenon known as "fast ice bridging." Each winter, he noted, there always seemed to be a location somewhere along Lancaster Sound where a barrier (or bridge) of landfast ice formed—sometimes earlier in the winter, sometimes later—which then served as a barricade to effectively halt the flow of Arctic sea ice as it moved down the channel toward Baffin Bay.

"The position of the Lancaster Sound ice bridge can vary dramatically year to year," Marko explained, "depending on air temperatures and other factors having to do with fast ice formation, with the two extremes being as far as eight hundred kilometers apart. Field biologists have known for some time about this variation. They've also known that it is important biologically, as it can lead to significant variation in bird nesting, seal whelping, and any number of other animal behaviors. Then why, I began to wonder, shouldn't it also be important climatologically?"

A simple computation based on average volumes of ice production in Lancaster Sound indicated that, in Marko's words, "the two extreme eastern and western positions of the ice edge could correspond to a seventy-five-cubic-kilometer net difference in total exported ice volume." When such volume is added to the ice advected by similar shifts in Smith and Jones Sounds, the total volume of ice to be added to the Baffin Bay sea ice cover in a warm year or subtracted in a cold year might be large enough to have a significant effect on later downstream ice behaviors—perhaps even to explain the out-of-phase behavior of the entire region.

The one serious problem with Marko's hypothesis, as Marko himself was aware, was the small amount of hard scientific evidence to support it. At the time of the 1991 study there were almost no available data on the specific volumes of ice advected through the side channels, no tracking of the ice as it moved across Baffin Bay, no measurements of how it might be affecting the ice farther downstream. Most of the evidence that did exist was circumstantial, based on correlations between the timing and location of ice edge stabilization in Lancaster Sound as compared with indices of later ice extents in the Davis Strait.

If these Lancaster Sound correlations had been the only evidence in support of Marko's hypothesis, the idea might never have gained currency among the forecasters at the Canadian Ice Centre, and I might never have heard about it during my contacts with them in the summer of 1991. But about the same time the Marko study was being produced, another dramatic sequence of ice events was occurring in Smith Sound, at the northernmost extremity of Baffin Bay—events that added a great deal of credibility to Marko's proposition.

Just as in the other Baffin side channels, a bridge of fast ice typically forms across the northern part of Smith Sound in early winter, blocking the flow of multiyear ice from farther north. But unlike the bridges that form in the other channels, this one is particularly evident to methods of aerial observation, as it also defines the northern edge of a large and persistent area of ice-free water known as the North Water polynya.

Beginning in 1973, as a matter of routine, forecasters at the Canadian Ice Centre had started tracking the annual position and timing of the Smith Sound ice bridge and the subsequent definition of the North Water polynya using infrared imagery from circumpolar satellites. According to their observations, the bridge had formed normally in December or January for seventeen years, from 1973 through 1989. Then in 1990, for the first time since record keeping started, the bridge failed to form. A year later, in 1991, it failed to form again until April, and then it appeared only for a brief time and much farther north.

*Brendan* last sailed to Labrador in 1991—the year of the heavy ice—and at least for forecasters at the Canadian Ice Centre, the correlation between the failed Smith Sound ice bridge and the anomalous ice extents in the Labrador Sea had become increasingly difficult to ignore. Even though this new evidence was still only circumstantial, it tied in perfectly with Marko's explanation for the out-of-phase behavior of this region, and its timing was exactly as Marko himself might have predicted.

The only trouble was that there were still no actual measurements to verify the connection. In this regard, the situation in Baffin Bay was very much like the situation Parkinson had observed almost twenty years earlier in the Sea of Okhotsk. In both cases a pair of events appeared to be related. In Parkinson's case it was the heavy ice year combined with the northerly winds from Siberia. In the present case it was the failure of the ice bridge in Smith Sound combined with the heavy ice along the Labrador coast.

The potential correlations were there, staring observers in the face. And the logic was there, providing plenty of reason for suspecting a cause-and-effect relationship. But the same thing that had happened with Parkinson in the Sea of Okhotsk could also happen with observers of the Baffin side channels. Years from now, after measurements had been collected and the data set had been lengthened, the correlations that had once appeared so strong and the logic that had seemed so convincing might both turn out to be pure coincidence.

Marko himself makes a similar point in his two later studies, especially in the one published in manuscript form in May 1994, about a month before *Brendan* sailed. In this study—a gridded forecasting model for LDB sea ice cover—Marko retreats from his earlier emphases on the importance of side channel ice exports. The problem, he explains, lies not in the compelling logic of the idea but in the difficulty of assembling hard scientific evidence, one way or the other.

So what does all this mean . . . that Marko is wrong? That the side channel export hypothesis gets shoved into a footnote somewhere and forgotten? Not quite. What it means is what Marko himself says it means—that the hypothesis is tentative, speculative, in need of more study. It is an idea that is logical and that rather neatly explains the out-of-phase behavior of the downstream ice. But it's also an idea about a region that has been only superficially investigated. The correlations that Marko describes in Lancaster Sound, the correlations others have noted in Smith Sound—both may emerge someday as indicators that the hypothesis works, or both may turn out to be coincidence. All anybody can say at this point is that the jury is still out. We simply lack the hard evidence to make a call.

*July 8, Friday afternoon. Chateau Bay, Labrador.*

Mike squats in the stern of the rubber boat this afternoon and operates the outboard engine for the half-mile run out to Castle Island. He steers toward a landing place on the western shore, guns the engine at the top of a swell, and rides the crest up the incline of the beach. As the bow touches the sand, all five of us jump out and carry the boat to a row of boulders where it can be tied and left for a time out of harm's way.

As soon as we leave the beach, Blue turns south and heads out alone across a rubbly badland. Pete squats where he is and begins exploring a miniature world of mosses and lichens and tundra flowers. Mike and Amanda and I continue toward the center of the island, picking our way among fallen boulders until we come to a section of broken rock at the edge of the basaltic wall that I know will take us to the summit.

With Amanda in the lead, we ascend a fluted staircase, moving from handhold to handhold until we emerge at the top and find ourselves staring across a shelf of barren rock. Then, as Amanda and Mike set out to explore the perimeter of the shelf, I move directly across to a place at the eastern edge where I'd stood three years ago when I'd come here to survey the ice.

I have a funny feeling about this place—a feeling that there should be no time here. In some ways this feeling proves accurate. The shelf where I'm standing is the same as it has been for thousands of years—a blasted moonscape without vegetation, without variety or contour. But the vistas that surround it—the hills, the coves, the bays, the strait itself—have all changed dramatically.

During *Brendan*'s last visit here, in August 1991, winter was still apparent everywhere. There was snow on the hillsides and in the coves and along the beaches. Chateau Bay was clogged with ice. Near the mouth of the cove where *Brendan* was anchored was a large tabular ice floe—rough and hummocky in appearance, bluish in color, fresh to the taste. Another floe of similar description blocked the harbor entrance, while a third lay grounded near the landing place where we'd left the dinghy. Out in the strait itself the water was close to ten percent ice-covered, with pack ice streaming in rows and icebergs everywhere. The water temperature was close to freezing, and the air temperature wasn't much warmer.

In contrast, the water in the strait this year is five or six degrees Celsius and the air this afternoon is near ten degrees—a Labrador heat wave! There are only a few snowy patches left on the north-facing hills and only twenty-seven bergs visible by actual count in the strait, with no icebergs in the bay and no sea ice anywhere.

Ice, I remind myself, is a climatic accelerator. More ice wants more ice. Less ice wants less ice. Each situation tends to amplify the condition that already prevails. It is hard to imagine another kind of climate where the conditions could vary so dramatically during the same season and in the same location from one year to the next. And we are here this year nearly a month earlier than in 1991!

I gaze out toward Belle Isle and let my mind wander, thinking about ice riddles, thinking about John Marko and the side channel export hypothesis. I realize that at the end of my encounter with Marko and his work I had felt frustrated. First I'd explored Marko's idea and had learned of its compelling logic. Then, along with Marko, I'd concluded that the whole proposition was circumstantial and I'd been forced to relegate it to the status of an interesting but unproven idea—not the unambiguous answer I was hoping to find but simply one more part of a complicated question.

One consequence of tabling Marko's hypothesis was that the door had been thrown open again to fresh speculation. The "east Canadian/west Greenland cold spot" hadn't gone away. The out-of-phase behavior of the LDB ice regime hadn't changed. The "brutal fact" of this anomalous area was still here. Only the explanation as to why it was here seemed as far away as ever.

The next step in the investigation for me began a little over a year later. I'd gone to Washington for a follow-up meeting with Claire Parkinson—specifically to talk about the Marko hypothesis and to ask about other mechanisms that might be causing the cold spot anomaly. She listened patiently. Then she did two things. First, she asked how much I knew about "thermohaline circulation"—ocean currents—and particularly about "deep-water formation" and a global phenomenon known as the Great Ocean Conveyor Belt. And second, she mentioned the name of a well-known teacher and researcher working out of Columbia University's Lamont-Doherty Geophysical Observatory, a geochemist named Wally Broecker.

"Read some of Broecker's articles," Parkinson suggested. "Then go talk with him. Tell him about John Marko's idea and see how he reacts. Ask him what he thinks is going on in the Labrador Sea area. Maybe he'll have some ideas for you."

Although I couldn't have known at the time, this series of offhand suggestions eventually defined a new phase in my investigation. They pointed not only to Lamont-Doherty and Wally Broecker but also to an entirely new set of concerns: away from the LDB area and into the deep ocean, away from the circulation of the atmosphere and into a dynamic process called "deep convection" and an engine known as the North Atlantic Deep Water that most scientists now regard as the power plant that drives the circulation of the global ocean.

When Amanda and Mike return from their circumnavigation of the cap, I'm still staring out at the strait in a half reverie, pondering the intricacies of ocean and air and ice. Amanda sits down cross-legged at the edge of the overlook while Mike stands next to me, both of them gazing off to the north.

Nobody says anything for a time, but for some reason I know what both are thinking. Maybe it's the way Mike sniffs the air as he looks out past Belle Isle. Maybe it's the way Amanda stares with a kind of mute expectancy toward the next headland—and the next, and the next. An invisible attractor out there is pulling at their minds—pulling at all our minds—and I know what that attractor is.

"How about it," I say to Mike. "Are we ready to try for Greenland?"

"There's a storm system crossing Hudson Bay tonight," says Mike. "Northwest gales forecast for the Davis Strait tomorrow."

"Maybe we'll just follow the coast for a couple more days, then."

"What about the icebergs?" says Amanda. "We can't sail along this coast at night."

"The nights are getting shorter with every kilometer," I remind her. "But you're right. Until we get into the Arctic daylight, we'll have to anchor—or throttle down if there's no place to stop."

The three of us become silent again. I try to picture the next leg of our journey in my mind's eye—a thousand kilometers due north, first following the Labrador coast, then crossing Davis Strait obliquely from somewhere near Hamilton Inlet, Labrador, to the headlands of the Gotthabsfiord in western Greenland. A thousand kilometers through ice, through all three North Atlantic storm tracks, through a section of ocean that is among the loneliest in the Northern Hemisphere. And at the conclusion of that passage a desire, a promise, a commitment, a goal that each one of this crew has now made his own: Greenland.

# IV.

# North

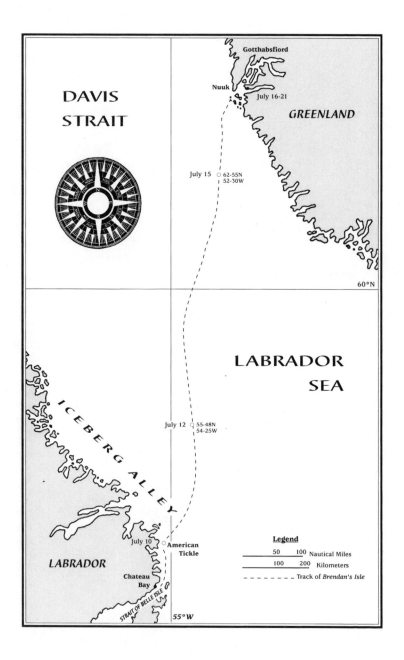

DAVIS STRAIT

Gotthabsfiord

Nuuk

July 16-21

GREENLAND

July 15 ○ 62-55N
52-30W

60°N

LABRADOR

SEA

ICEBERG ALLEY

July 12 ○ 55-48N
54-25W

July 10 ○ American
Tickle

LABRADOR

Chateau
Bay

STRAIT OF BELLE ISLE

55°W

Legend

50        100 Nautical Miles

100        200  Kilometers

– – – – – Track of *Brendan's Isle*

*July 12, Tuesday morning. 55/48N, 54/25W.*

*LOG: Wind: west, 6 knots. Barometer: 29.05 steady. Weather: overcast, fog, cold. Position: Labrador trench, 130 miles E of Cape Makkovik, Labrador.*

At four o'clock, Blue and I take over the deck in thick fog. *Brendan* is under power, with the engine ticking over at a low rpm and the boat making about three knots through the water. My watchmate and I are having a problem again, as we always seem to do when the engine is running. We do not argue this time—we just maintain a stubborn silence and avoid each other's eyes every time we exchange places, moving from the helm to the bow watch and back to the helm again every half hour.

We're looking for ice this morning, throttled down so the person on the bow has a few seconds to warn the helmsperson about which way to turn. We're nearly through the Labrador Current, so the likelihood of meeting with iceberg ice is growing less with every hour. But according to a recent report, there are several bands of sea ice about thirty kilometers to windward. Since sea ice moves with the wind, we know these bands could be significantly closer—which means we could be sailing into the ice right now.

On a clear day when the helmsperson is able to see a few kilometers ahead, such occasional encounters with ice would present little problem. But this morning, with less than a hundred meters visibility, a sudden meeting with a large floating slab of sea ice might prove a little more dicey.

The light grows slowly under a heavy overcast. Finally, several hours into our watch, a west wind begins to raise a chop on the water. Blue and I initiate a welcome truce as I shut down the engine and invite him to take the next half hour at the helm while I move forward to watch at the bow.

It is cold now. Both water and air temperatures are only a few degrees above freezing—and in spite of the wool cap and ski gloves and heavy orange Mustang exposure suit that each of us wears while on deck, this bow-watching business becomes difficult after a time. The reflexes slow down. The brain grows numb. The eyes go a little blurry. And with everything out here looking the same shade of gray, it's easy to begin to imagine things.

Over and over I'm tempted to call out to Blue: "Watch out for that shadowy object off to the left . . . that area of brightness ahead . . . that dark loom to the right." But each time before I call out, I make myself wait a few seconds. I open my eyes wide and gaze ahead, and I watch the spectral shape that I thought was there slowly dissolve back into the gray.

After half an hour like this, I grow complacent. At one point I actually begin to nod off to sleep. Then without warning a flat white shape materializes a hundred meters ahead and doesn't go away. I stare at it and rub my eyes. As I do so, a second, larger shape appears, and a third. "Watch out, Blue! Ahead on the left! Do you see them?"

The boat swings sharply, and in a matter of seconds the shapes disappear back into the fog. Were they small iceberg pieces? Rotting floes of sea ice? It's hard to say. But one thing is for certain—they were *not* imaginary. I feel my pulse quicken and my breath come faster, and I'm suddenly wide awake again.

The prearranged time for contacting *Bowdoin* on the single-sideband radio comes soon after the change of watch. I sit down at the nav-station and begin to search the six-megahertz band for a working channel. Back in the lower latitudes this band was always full of traffic. Here, however, there is almost nothing. One channel has a pair of fishermen speaking in Danish. Another contains a conversation in Russian. The rest are empty, and I realize what a distance we've sailed and how far we've come from other people.

There are no other ships out here, possibly for hundreds of kilometers. There are no sounds of aircraft engines overhead. The only coast guard station along the east Labrador coast is at Cartwright, two hundred fifty kilometers astern, and that is little more than a lifeboat station. And the nearest settlement in Greenland is still almost eight hundred kilometers ahead.

I move to a calling channel and state *Bowdoin*'s name several times. No response. I move to another channel and try again. No response. Back to the first. No response. Normally, a call such as this would end here. Clearly, one of our stations is not propagating a strong enough signal. But for some reason, this morning I keep trying. We are social creatures, after all, and I feel a need for contact.

After a second unsuccessful round of attempts, I sit and stare at the radio, thinking about our solitude, wondering what odd collection of motives must combine to move a group of people to sail into this kind of wilderness, so far away from the safety net of interdependencies that our society provides.

Suddenly I hear a voice—or a group of voices—that seem to be talking in a strange language. I lean nearer the set, place my ear next to the speaker, turn up the volume. It's only then that I realize that the sound is not coming from the radio at all. It's coming from everywhere else—from the right and left and underneath—coming from outside the boat, directly through the hull.

I zip up my Mustang suit, pull on cap and gloves, climb up the companionway to the cockpit. Amanda is steering while Mike is forward at the bow, both of them watching ahead for ice, so neither is yet aware of what is happening in the water behind them.

Black bodies, perhaps fifty of them, swim in random groupings across the horizon astern. They hang back, none approaching closer than about twenty-five meters, and they gallop through the water like dolphins. Even at this distance, however, it is apparent that most are two or three times the size of dolphins. They are jet black, with large pot-shaped heads and hook-shaped dorsal fins. They swim with an undulating motion, exposing most of their heads and half their bodies each time they rise for a breath. I have no question as I gaze past Amanda what creatures have come to visit us this morning: these are several large family groups of *Globicephala melaena,* the North Atlantic long-finned pilot whale.

The west wind has increased, and the boat now makes a steady hissing noise as it moves through the water. Out here the whale signals that I'd heard through the hull are no longer audible. Instead, there is another, softer sound that competes with our own—the quick, staccato rhythm of fifty pilot whales breathing.

Amanda looks quizzically at me, still unaware of what I'm staring at. I tell her I've been on the radio, searching around for a little intelligent conversation. "And look who's answered," I say, gesturing astern.

She turns around and gazes at the whales. A smile grows across her face as several of the nearest ones begin to accelerate and approach closer to the sailboat.

*July 15, Friday morning. 62/55N, 52/30W.*

*LOG: Wind: south, 30 knots. Barometer: 28.92 falling. Weather: overcast, gale warning tonight. Position: Davis Strait, 80 miles S of Gotthabsfiord, Greenland.*

For three days the watches fade into one another, the wind draws astern, and *Brendan* and her crew roll effortlessly toward Greenland. The weather changes every hour, with blue skies giving way to gray, spitting rain, then clearing again. There's been no fog since we left the Labrador Current, and we've spotted only one large iceberg. The nights have grown progressively shorter, so that by last night there were only two hours of dusky twilight.

The pilot whales no longer follow *Brendan* this morning, although there has been a constant parade of them during the past two days. On Wednesday in particular, they followed in large numbers, until it seemed she had become a kind of rallying point for every migrating pilot whale in the Davis Strait.

Last night the weather started closing in. Now, as the wind and seas build, *Brendan* lurches onto her beam ends. The drawers and cupboards down in the cabins erupt in a series of explosive crashes, while underneath the hull the rudder begins to cavitate, setting up a harmonic vibration that rattles the chimneys of the oil lamps.

On deck, Mike stands behind the helm and eases the boat down following seas. Pete and Amanda move forward to tuck a reef in the mainsail and stabilize the boom with a preventer so that it will not jibe accidentally as the boat skids downhill. I climb to the cockpit, clip my safety tether onto one of the jack lines, and gaze ahead at a wall of cumulus cloud that grows along the eastern horizon where I know the coast of Greenland will eventually appear.

Once the mainsail is secure, Pete and Amanda move back to the cockpit and wedge themselves between Mike and me. Meanwhile, I continue to stare ahead, watching the line of clouds that has been growing along the horizon. I'm about to mention something to my shipmates about the strange shapes in the tops of these clouds and the odd pattern of dark and light at their base. When I open my mouth to speak, however, the sounds that come out are not the words I'd planned to say but a kind of strangled expletive instead.

My god!—these shapes I've been staring at—they're not clouds at all. They are a wild, improbable series of jagged mountain peaks, ascending through swirls of cloud, crowned with horizontal slabs of pure white glacial ice. They are the coast of Greenland!

I stop trying to speak and only continue to stare. One by one, the others follow my eyes, gaze into the east, and realize what they are looking at.

The scale of this coast is massive. The shapes are surreal. The surfaces are frozen and colorless. I find myself wondering where it is on this planet that we've come. Is this a place where people live? Is there soil here? Is there vegetation? Is there safe haven somewhere in all this rock where we can hide from the wind? Is there shelter from these steep gray seas?

I close my eyes, trying to put this moment into some kind of perspective. But it's no use. There is no relief from the vast, impersonal scale of what I'm seeing. Greenland! The thought that we have sailed here in this little boat actually takes my breath away.

*July 16, Saturday. Nuuk, Greenland.*

Even when you're moored in a perfectly protected harbor, you can feel the anger of a Greenland gale. The rigging vibrates. The boat heels from one side to the other in the swirling wind. The top of the stove chimney rattles with the sound of the rain driving against it. Down below, the chimney hisses as the water leaking through a deck fitting runs down the pipe and vaporizes into steam.

*Brendan* lies rafted alongside two other vessels this morning—an old Danish fishing trawler and a small American sloop—at the center of a long wooden quay. The three rafted vessels strain at their mooring lines while behind them the rain drives against a dark cliff and cascades down the rock like a waterfall. Standing puddles—some like small lakes—fill the roadways around the harbor and cover the loading docks on the commercial quay.

Mike and the others on *Brendan*'s crew have chosen to brave the weather this morning and hike into the town center of Nuuk, several kilometers away. I've decided to take the day off and remain warm and dry in the cabin until this gale blows itself out. As soon as the others leave, I take a moment to finish a few items of ship's business. Then I move to the main saloon, pull out a manila folder marked "Wally Broecker," and settle down next to the diesel heater for a continuing encounter with the ice riddle, this time in the context of an important and far-ranging theory about the thermohaline circulation of the global ocean: a theory Broecker has named the Great Ocean Conveyor Belt.

First proposed in a series of scientific articles in the mid-1980s, Broecker's description of the Great Ocean Conveyor Belt has provided a key to the scientific investigation of Earth climate for nearly a decade. The idea centers around the notion of a planetwide system of ocean currents that carry heat and cold, salt- and freshwater to every corner of the world ocean. In addition to being a circulatory mechanism, this system incorporates a triggering device for delivering the system from one stable mode of operation to another. As such, it provides at least part of the answer as to why the global climate has undergone sudden, catastrophic changes in the remote past and why it may do so again at some undisclosed time in the future.

One of the first questions I found myself asking a year ago as I began reading about Broecker's idea was what in the world it had to do with ice. Why was I looking here, at the writings of a geochemist theorizing about the circulation of the global ocean? What was the connection? What did ocean currents have to do with increasing sea ice cover in the LDB area, for instance? Or with the east Canadian/west Greenland cold spot?

The answer, I discovered, has to do with a fundamental relationship. In its simplest sense, global climate is a dynamic circulatory system for distributing heat and chemical substances around the Earth. In reality, however, it is two parallel systems, one comprising gas (the atmosphere), the other comprising liquid (the oceans). The two systems may at first seem to operate independently of one another. Yet as climatologists have long realized, they are integrally linked, and it is impossible to understand the behavior of one without at the same time working to understand the behavior of the other.

I had already explored the connection between sea ice and the atmosphere in the context of John Marko's side channel export hypothesis. Next I needed to explore the connection between sea ice and ocean currents—a connection that had to exist, I realized, or *Brendan* could not be where she is today: in an ice-free harbor on an ice-free coast only a few hundred kilometers south of the Arctic Circle.

Here in western Greenland we've managed to sail our way into an apparently impossible situation—a huge geographical cul-de-sac of ice. Just beyond the mountains to the east is the largest glacial ice sheet in the Northern Hemisphere. A few kilometers to the west is the edge of the Baffin ice pack, and beyond that is a frozen sea extending all the way to Baffin Island. To the north is another area of frozen sea that reaches to Ellesmere Island and the North Water. And to the south is a tail of sea ice that curls around the southern tip of Greenland and clogs the fiords of the southwest coast until late in the summer.

Lying in Nuuk harbor, we are surrounded in nearly every direction by ice. Yet along this coast there is only an occasional iceberg visible and there is seldom sea ice anywhere—either in summer or in winter.

The mechanism for this strange state of affairs, as I'd seen many times in the ice atlases, is an extension of the drift system of the Gulf Stream called the West Greenland Current. As this current flows northward along the Greenland coast, it transports heat and salt in the upper layers of the ocean, and together these inhibit the formation of sea ice, sometimes for hundreds of kilometers north of the Arctic Circle. Along with this influence (and partly because of it) the current also affects biological populations, air temperatures, and numerous other components of the west Greenland environment.

So here is the answer to my question: ocean surface currents are importantly related to a broad range of climate features, including air temperature anomalies and the formation and distribution of sea ice. But having established this relationship, I found myself asking even more questions. How do large-scale ocean currents operate? What are the forces that drive them? What are the triggering devices that cause them to speed up or slow down, turn off or turn on? How do they respond to large human-generated changes, such as ozone depletion or $CO_2$ doubling, that are currently taking place in the climate system as a whole?

As luck would have it, Columbia University's Newberry Professor of Geology, Wallace S. (Wally) Broecker, has been asking these same questions for more than four decades. As a teacher, Broecker has been challenging generations of Columbia students with his provocative and far-ranging scientific ideas about the great geophysical engines of Earth climate. As a research scientist, using techniques he's helped to develop, he has been working since the mid-1950s on the measurement and tracking of chemical isotopes—in the atmosphere and in ocean bottom sediments—as a means of investigating some of the telltale fingerprints of Earth climate, both present and past.

But Wally Broecker is more than just a geochemist and more than just a climatologist. As Richard Fairbanks, one of his scientific colleagues at Columbia, has described him, he is "the grandmaster of global thinking. He just keeps bridging fields—geochemistry, oceanography, paleoclimatology—to home in on the big picture."

"He's provided the gems of ideas and guidance for entire fields of study," Fairbanks said at another point. "His intellectual guidance and intuition as to what is important is driving hundreds of scientists today. Hundreds of scientific initiatives around the world are tracking down his ideas."

Broecker has been recognized many times by his peers for the importance of his scientific contributions. In 1979 he was elected to the National Academy of Sciences, and soon afterward he was awarded the Vetlesen Medal (the equivalent in the Earth sciences to the Nobel Prize) for his work on the use of chemical tracers to investigate patterns of global thermohaline circulation. Just as Maurice Ewing had been a quarter century earlier, so Wally Broecker has today become one of the mythical "great men" of Columbia's Lamont-Doherty Geophysical Observatory. Recently, when an eminent foreign scientific visitor was shown the Lamont-Doherty geochemistry building, the visitor exclaimed, "Oh, so that's Wally Broecker, Inc."

Probably the most important idea to have come out of "Wally Broecker, Inc." during the past decade has been the Great Ocean Conveyor Belt—an idea that had a typically Broeckeresque beginning. As a man who believes profoundly in the creative process, Broecker has always enjoyed the tension that evolves out of apparently discontinuous subject matter. "I like to work with very different types of information," he says. "You have ten sources of evidence and they all don't quite fit together. That's why nobody understands what's going on. So you think about it and you say, 'Which one of these things am I really going to put my money on? . . . Which am I going to take seriously?' "

In the case of the Great Conveyor Belt, the ideas he decided to take seriously evolved from at least three inputs. First was the work Broecker himself had been pursuing for several decades tracing the routes and dating the ages of deep waters in the Atlantic and elsewhere by means of radioactive isotopes. This provided the basic understanding of oceanic circulation and the conceptual framework against which to evaluate the other inputs.

Then came a new set of paleoclimatic records of the last ice age as revealed in a glacial ice core drilled at the Dye-3 site in southern Greenland in 1980–81. An earlier core drilled at Camp Century in northern Greenland in the late 1960s had shown that the ice age climate had tended to jump around a lot, with periods of intense cold shifting suddenly into warmer interludes, then shifting back to cold again. The new core in southern Greenland revealed the same pattern, and when chemical analysis of the air trapped in the ice showed that the shifts were always accompanied by large changes in atmospheric $CO_2$, Broecker realized that there had to be a strong air–ocean connection.

The third input came at a lecture that Broecker attended in Bern, Switzerland, in 1984. In this lecture Swiss oceanographer Hans Oeschger talked about a strange, quantum-like behavior that seemed to characterize the oceanic circulation system during the last glacial period, with the system tending to flip back and forth between two stable states. These states, Oeschger suggested, corresponded to the sudden climatic shifts between warm and cold periods in the ice core record, suggesting some kind of oceanic on/off switch.

Several components of the Great Conveyor Belt hypothesis were now in place: a planetwide mechanism for thermohaline circulation, a record of sudden episodes of climate change from the geologic past, and an educated guess that the circulation system somehow had the capacity to flip on and off. What remained for Broecker was to tie the whole business together, first by providing an answer for how the on/off modalities correlated with the abrupt climate changes observed in the ice core record, and second by suggesting a triggering mechanism and explaining how it worked.

Perhaps the easiest way for a nonscientist to visualize Broecker's Great Conveyor Belt is by means of a diagram first published in 1987 in *Natural History* magazine [see the next page]. The diagram is admittedly oversimplified. Broecker, in fact, calls it a "cartoon" and emphasizes that it is merely a representation of a much more complex series of circulation loops and pathways that exist in the real ocean-atmosphere system. Nevertheless, the diagram provides a dramatic image of the Conveyor Belt concept and, in Broecker's words, "symbolizes the importance of linkages between realms of the Earth's climate system."

As the diagram suggests, the Great Conveyor Belt is a system of deep and near-surface currents that flow up and down the Atlantic, Indian, and Pacific Oceans and around the Antarctic continent. The main energy source and the point of origin for the lower limb is an area in the northern Atlantic (marked "Sea-to-Air Transfer" in the figure) where highly saline surface waters are cooled through contact with frigid air from the Canadian Arctic. The cold water increases in density until it sinks into the abyss and begins to flow southward, forming what oceanographers call the North Atlantic Deep Water (NADW), a "sluggish mass that fills most of the deep Atlantic."

The NADW is the engine that drives all the rest of the system. It surfaces briefly in convective loops that form in the Southern Ocean around Antarctica. Here it is super-cooled and sent tumbling back down into the abyss where it forms the Antarctic Bottom Water (AABW), the coldest and densest water in the oceans. Tongues of this water flow north into the Pacific and Indian Oceans where they warm and

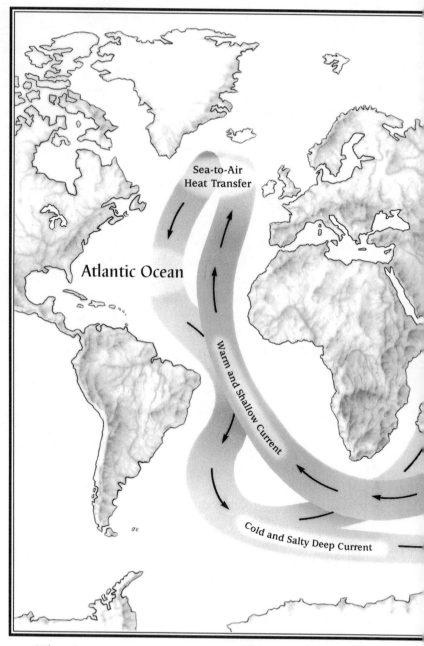

Sea-to-Air
Heat Transfer

Atlantic Ocean

Warm and Shallow Current

Cold and Salty Deep Current

**The Great Ocean Conveyor Belt: A worldwide system o**

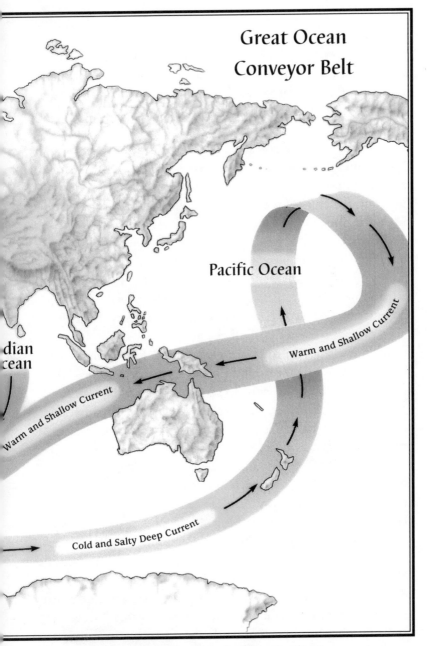

Great Ocean
Conveyor Belt

Pacific Ocean

Indian
Ocean

Warm and Shallow Current

Warm and Shallow Current

Cold and Salty Deep Current

oceanic currents with global climatic impact.

rise to the surface. Here they become a complex system of near-surface currents that eventually reenter the Atlantic, move north, and spread back into the area where they can start to cool and thus begin the process all over again.

In an interview several years ago, Broecker commented on the "stupendous magnitude" of the Conveyor Belt system. "I'm going to use a unit called a Sverdrup," he said, "named after a famous Scandinavian scientist. A Sverdrup is one million cubic meters of water per second. That's sort of a meaningless number because it's so big . . . all the rivers in the world carry one Sverdrup of water. So if you put all the rivers in the world in one big pipe, out the end would come one million cubic meters per second. All the rain in the world falls at the rate of about fifteen to sixteen Sverdrups. The Conveyor Belt flows at a rate of twenty Sverdrups. Twenty times the flow of all the world's rivers!"

Broecker goes on to explain why this huge oceanic circulatory system is so important to the Earth's climate. "The [warm surface] water enters the northern Atlantic at about 10 degrees Celsius and sinks to the bottom at a temperature of about 3 degrees Celsius. So every cubic centimeter of water that makes this loop gives off 7 calories of heat to the atmosphere. When you add that up over a whole year, it's a staggering amount of energy. It's equal to something like a third of the solar energy that reaches the surface of the Atlantic north of 35 degrees North!"

It's easy to understand, when you think about these numbers, why the operation of the Conveyor Belt is so important to the climate system. Northern Europe is the first and most direct recipient of the excess heat that is transported north into the Atlantic. "The amount of heat [Europe] receives is jacked up by thirty percent by the heat that comes out of the ocean," Broecker explains. "If the Great Conveyor Belt were to shut down, it would get so cold in northern Europe during the winter that no trees could grow."

One of the most persistent questions that Broecker is asked about the operation of the Great Conveyor Belt has to do with this notion of shutdown. If, as Broecker proposes, the ocean circulatory system has at least two modalities (on/Conveyor and off/Conveyor), and if these two modalities correspond to radical global shifts between temperate and ice age climates, then how does the system move from one modality to the other and what is the switching mechanism?

The answer to this question—in a word—is salt. But to understand this answer, you have to go back to the place where the Conveyor Belt begins, the northern Atlantic, and look at why the formation of a deep-water layer can take place in this particular part of the world ocean and nowhere else.

In order for water at the surface of the ocean to sink to the bottom, it must be heavier (more dense) than the water underneath it. Cold water is more dense than warm water; salty water is more dense than freshwater. Thus, any place in the ocean where the surface is cooled by frigid air masses blowing across it and where the salt content is allowed to build up, the water will become progressively more dense until it finally begins to sink.

The average salinity of ocean water worldwide is thirty-five grams per liter. But because of water vapor loss to the atmosphere, the average salinity of the surface water in the northern Atlantic is thirty-six grams per liter. Since one gram per liter of salt is equal in its density effects to four degrees Celsius of cooling, the surface water in the northern Atlantic is the equivalent of four degrees colder than the surface water any place else in the world—even before it starts being cooled!

Broecker elaborates: "In the North Pacific, where the salinity is low, you can cool water to its freezing point and it only sinks a couple hundred meters. It doesn't have a high enough density to go down farther. The waters underneath it are more dense. In the North Atlantic, if you cool water to plus-two degrees Celsius, it'll go to the bottom, flow down the Atlantic, and get out the other end."

The reason the Atlantic is saltier than other oceans has to do with a combination of global weather patterns and geography. The prevailing westerly winds that blow around the Earth in the north temperate latitudes carry water vapor away from the Atlantic and deposit it in areas where the runoff also flows away from the Atlantic. This situation results in a net reduction of freshwater and a net buildup of salt that increases the density of the surface layer and drives the Great Conveyor Belt.

As Broecker is quick to point out, there is an Achilles heel in this system, however. If for any reason the surface layer of the northern Atlantic were to receive a sudden influx of freshwater large enough to reduce the average salinity by one or two grams per liter (a condition described by climatologists as the "halocline catastrophe"), deep convection and the formation of North Atlantic Deep Water would cease, and the whole operation of the Conveyor Belt would shut down.

In fact, there are a number of ways that such a sudden influx of freshwater could take place. Very large amounts of precipitation occur in the polar and subpolar regions and in the temperate areas that adjoin them. Thus, rain, snow, and continental runoff pose a constant threat to the Conveyor system. So does any event resulting in the export of large amounts of freshwater ice into the northern Atlantic—the abrupt slipping of a large glacial ice sheet, for instance, or a pulse of multiyear sea ice migrating southward from the Arctic Ocean.

The result, says Broecker, is "a system with multiple identities, capable of undergoing major reorganizations." The on/off switch is what he calls a "salt oscillator," with the on/ Conveyor mode corresponding to periods of relatively high salinity in the surface layer, and the off/Conveyor mode corresponding to periods of relative freshening.

The evidence that such an on/off switch exists is everywhere apparent in the Greenland ice core records. In fact, since Broecker devised his hypothesis and began describing the Great Conveyor Belt, two more ice cores have been drilled. The Greenland Ice Core Project (GRIP) core, drilled by a team of European scientists, has now been extensively analyzed and confirms the same pattern of abrupt, sometimes decadal shifts between ice age conditions and warm conditions during the last period of glaciation. And to complete the picture, a series of major freshening episodes, called "Heinrich events," have now also been proposed and convincingly demonstrated, and their timing has been shown to correspond well with the abrupt on/off timing of the ice age climate changes.

What all this means is that over the nine or ten years since Broecker first began to describe his Great Conveyor Belt hypothesis, a good deal of additional evidence has been assembled, suggesting that the ocean-atmosphere system does indeed work in the way that Broecker has proposed.

The idea is grand and far-ranging, of course, and many details of its operation have been left for others to work out. As Richard Fairbanks has said, there are probably hundreds of scientific initiatives around the world dedicated to the testing, clarification, and elaboration of Wally Broecker's idea. Rather than diminishing it in any way, however, this flurry of scientific activity may be the sincerest testimony of all to its greatness.

*July 16, Saturday noon. Nuuk, Greenland.*

I break from my deliberations for a time as the lunch hour approaches. The wind still groans in the rigging, and the boat heels sharply with every fresh gust that ricochets off the cliffs above. I pull on my Mustang suit and climb to the deck to inspect the mooring lines. Then I make myself a cup of soup and slip into the nav-seat to check the various instruments and radios—a kind of knee-jerk reaction, as if *Brendan* were still under way somewhere far out in the Davis Strait.

I glance at the compass, the wind speed indicator, the barometer. I flip on the single-sideband radio and listen for a moment to the stony silence caused by the high land around us. Suddenly I find myself thinking about *Bowdoin,* wondering where she is this afternoon, imagining her hove-to somewhere south of Cape Farewell, riding out this gale in what is surely the stormiest part of the North Atlantic.

I dial up to the six-meg calling channel and, even though I know there will be no response, I speak *Bowdoin*'s name several times into the microphone. The only sounds I hear

are the wind in *Brendan*'s rigging and the tinny echo of my
own voice as it modulates through the radio speaker. A sailing voyage is a lonely undertaking, I think, in spite of all the
electronic gadgetry that each of our vessels carries. Whatever
*Bowdoin* and her crew are doing out there, whatever difficul-
ties they may be encountering, they will just have to deal
with them on their own and trust to their own resources.

After finishing my soup, I shut down the radio and just sit for
a time, listening to the wind and rain. Soon I find my
thoughts returning to Wally Broecker, the Great Conveyor
Belt, and some of the ideas that I'd been reading and thinking about all morning.

Many students of Broecker's work (myself included), after
pondering the Conveyor Belt hypothesis, have found themselves asking how the Conveyor system may be relevant to
our own time period. Virtually all of Broecker's examples of
sudden changes between the on/Conveyor mode and the
off/Conveyor mode seem to come from the ice ages. Those
events are obviously important. But what about now?
What's going to happen to the Conveyor Belt as the concentration of carbon dioxide increases in the atmosphere, for
instance? Can the system shut down during a warm period
like the one we're living in—or does it turn on and off only
after events that take place during an ice age?

Broecker himself provides what may be the most honest
answer to this question. "Scientists don't yet have the foggi-
est idea whether what we are doing to our planet threatens
to change the behavior of the Conveyor. All we know at this
point is that we're giving the system the biggest jolt imagin-
able. The effect of $CO_2$ doubling is as big as a Milankovitch
cycle [one of the hypothesized causes of ice ages]—as big as
any of the great events that have precipitated radical climate
changes in the past."

There is a race going on, Broecker states, between researchers, who are struggling to comprehend how the Earth's climate system works, and humankind in general, who are perturbing this same system with our runaway population growth and our increasing use of fossil fuels. Broecker often talks about the "coming surprises" that we are sure to encounter as we continue to play a kind of climatological Russian roulette with our planet.

"There are those that believe that we live in some sort of God-given stable system that we're powerless to perturb," he says. "I don't believe that at all. The system we live in is capable of doing outrageous things. Thus when we hit our Earth system with greenhouse gases, perturb it, the range of possibilities of what might happen is quite large."

When Broecker is asked what some of these possibilities are, his speculations range over a broad spectrum. For example, the cover of the issue of *Natural History* magazine in which the Great Conveyor Belt article first appeared contains a sales stimulator that warns: "Europe beware: the big chill may be coming." Broecker was annoyed—because, in fact, no mention of the Conveyor's future had ever been made in the article. "The fact is, I thought, at that time, that the coming greenhouse warming would, if anything, strengthen the Conveyor by increasing the rate of vapor loss from the Atlantic basin."

Somewhat later, Broecker began to consider the opposite alternative: that greenhouse warming could lead, at least temporarily, to Conveyor shutdown. "In addition to increasing vapor export from the Atlantic basin, the greenhouse warming [might] increase the transport of fresh water to the northern Atlantic. In the short term (i.e. decades), the salinity decrease created in northern surface water [might] be more important than the Atlantic-wide salinity increase caused by increased vapor loss from the Atlantic basin."

The bottom line, as Broecker now emphasizes, is that

there are probably a number of stable states of operation in the Conveyor system in addition to the two he first hypothesized—states that we don't yet fully understand.

"We're dealing with a system that is very cleverly designed and has all kinds of funny things about it that can create non-linear responses," he says. "This should give us more respect for the system we're in the process of perturbing."

*July 21, Thursday. 64/13N, 52/42W.*

*Log: Wind: light northerly. Barometer: 29.47 steady. Weather: broken overcast, fog. Position: Fyllas Bank, en route toward Manitsoq, Greenland.*

The sun glows in a hazy circle through the fog as I take over the helm this morning at ten o'clock. *Brendan* has been under way since she left Nuuk harbor just after five, motoring through heavy fog with the mainsail hanging limp and the booms and spars dripping a steady rain onto the decks.

Now, as the sun begins to burn through the overcast, the boat feels as if it is floating on air. The fog is like velvet. The surface of the sea is perfectly smooth. Both sea and sky are the same silver-gray, and there is no line of definition between them. As we move forward through this featureless medium, I have to keep reminding myself to watch for small iceberg pieces that may not be visible to my watchmate on the radar below—and for tree-sized logs that float on the surface as they also move north in the current.

*Brendan* is still a little more than two hundred kilometers from the Arctic Circle this morning and several days' sail south of the midnight sun. But even here there will be no darkness—only a murky gray light that feels little different at midnight than it would on a cloudy afternoon.

The yellow circle of sun that has now emerged through the overcast hovers at spreader height perhaps thirty-five degrees above the horizon, on its way in a large, shallow circle around the perimeter of the sky. At six o'clock this morning it passed on a nearly horizontal course just above the mountains to the east. At noon it will cross our stern, due south. By late afternoon it will swing into the west and begin to drop imperceptibly toward the horizon. Then as midnight approaches, it will circle north and slice into the surface of the sea, there to hide a few degrees below the horizon for an hour or two before reemerging as another dawn.

I stare ahead, listening to the drone of the engine, thinking about the flurry of preparations my crew and I have made since the gale finally blew itself out yesterday. After the others returned from their walk into town, we spent the afternoon checking the hull and mechanical systems for wear, changing headsails, topping off the tanks with water and fuel. For an hour Mike and Blue hung on a bosun's chair in the rigging mounting a contraption called an "ice barrel"—a crow's-nest affair modeled after a similar item that Captain Mac had invented years ago and that *Bowdoin* carries on her foremast. *Brendan*'s ice barrel sits on her main spreader, twenty-five feet above the deck, where it will provide a safe platform for a lookout to con us through the ice when the going gets thick.

I smile to myself as I stare at the blank wall of fog ahead. We're ready now, I think, for the final leg of our northward dash—eight hundred kilometers along one of the wildest coasts in the world to the edge of the ice river in Disko Bay.

# V.

# Kangia: The Ice River

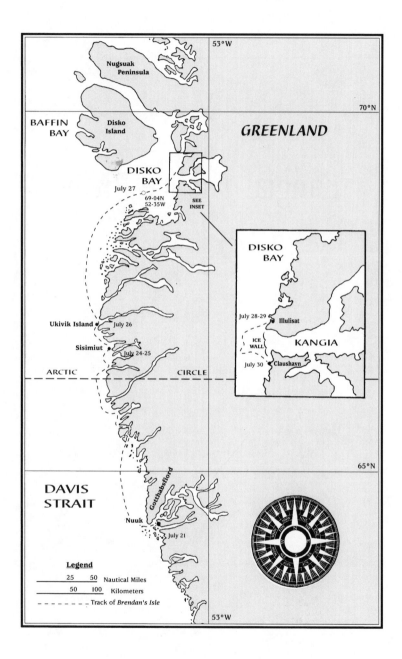

53°W

Nugsuak
Peninsula

70°N

BAFFIN
BAY

Disko
Island

GREENLAND

DISKO
BAY

July 27

69-04N
52-35W

SEE
INSET

DISKO
BAY

Ukivik Island    July 26

Sisimiut

July 24-25

July 28-29    Illulisat

ICE
WALL

KANGIA

July 30    Claushavn

ARCTIC

CIRCLE

65°N

DAVIS
STRAIT

Gotthabsfjord

Nuuk

July 21

Legend

25      50    Nautical Miles

50      100    Kilometers

－－－－－－    Track of *Brendan's Isle*

53°W

*July 25, Monday. Sisimiut, Greenland.*

Traveling along the coast of western Greenland in a little sailboat is an experience of opposites. Most of the route is lonely and wild. The fringe of islands along the immediate coast is barren and empty of vegetation. A few kilometers inland, black mountains rise to heights of two thousand meters, fronted in many places by steep foothills, crenelated by deep fiords, capped by the eerie white loom of the great central ice sheet.

For kilometer after kilometer as you travel along this coast, there are no signs of human habitation—no roads or buildings, no towers, no electrical transmission wires, nothing to suggest that a single human being has ever set foot upon this land. Then, just as you begin to think that you and your shipmates may be the only people left in the world, you notice a navigational light on an island ahead, a radio tower silhouetted against the sky, a cluster of geometric shapes framed against the rock. In a few more kilometers these shapes resolve themselves into the familiar patterns of buildings and rooftops, and you realize that you are approaching a modern settlement.

In fact, the entire west Greenland coast for nearly twelve hundred kilometers, from Qaqortoq in the south to Upernavik in the north, has been organized into a series of just such modern settlements. Originally established as Danish fishing stations, they have now evolved into regional centers of population, each consisting of between a thousand and five thousand inhabitants. They have been established along the coast at intervals of about two hundred kilometers, and although there are no roads connecting them, they are linked by coastal cargo ships and passenger ferries and by a regularly scheduled helicopter service.

The first settlement of this type that *Brendan* visited on her way north was Manitsoq, a village of about two thousand inhabitants nestled into a bowl of rock on an island about a hundred sixty kilometers north of Nuuk. The second is Sisimiut, a settlement of some five thousand inhabitants located on a long, narrow bay where we've been anchored since yesterday, about three hundred fifty kilometers north of Nuuk and fifty kilometers north of the Arctic Circle.

This afternoon shortly after two o'clock *Brendan* moves out of the anchorage area and proceeds into the channel in front of the fish plant pier. Moored alongside this pier is a large traditional schooner with a plain spoon bow and slab sides. Her twin masts and her black-tarred ratlines stand out in stark relief against a backdrop of commercial buildings. A large American ensign hangs from her taffrail, framing the name on her transom: *Bowdoin*.

It comes as no surprise to me that *Bowdoin* is in Sisimiut. I've been in regular radio contact with her skipper for the past week, and I've known since yesterday that she would be arriving here today.

*Brendan*'s crew sets up bumpers and mooring lines, and in

a few more minutes we are nested gunwale to gunwale alongside the schooner. I climb over her bulwarks and onto oiled pine decks, and I begin looking for her skipper. A young fellow with a light, neatly trimmed beard steps forward. I'm about to ask where I might find the ship's master when he holds out his hand in a gesture of welcome. "I'm Elliot," he says. "I feel like I'm finally getting to meet an old pen pal."

There is much news to exchange. We begin with a comparison of our journeys north, then we talk about the day *Brendan* spent moored in Nuuk harbor while *Bowdoin* was hove-to in a force-nine gale south of Cape Farewell. Elliot passes lightly over this event—it is, after all, a necessary part of life for a bluewater sailor—but I sense that some of his less experienced crew may not have been quite as resigned as their skipper to the inevitability of such an ordeal.

Finally, we come to talk about future cruise plans. I learn that *Bowdoin* intends to depart from Sisimiut for Disko Bay tomorrow noon—also that she plans to travel the entire three hundred kilometers in a single twenty-four-hour run. Elliot has already seen the ice fiord at Disko the year he sailed there as *Bowdoin*'s mate in 1991. His eyes grow wide as he tries to describe the place.

"You won't believe it until you see for yourself," he says. "It's one of the great natural wonders of the world. There's nothing else like it on the planet!"

One of my hopes for this summer's voyage has been to rendezvous with *Bowdoin* at the mouth of the Disko ice fiord. The chances seemed unlikely until now—but suddenly it appears that such a meeting may be possible only if *Brendan* and her crew are also willing to dash northward tomorrow for the final three hundred kilometers.

I withdraw to *Brendan*'s cabin for a powwow with Mike and the others. The decision takes only a moment. To a person, *Brendan*'s crew is eager to make the passage and to meet *Bowdoin* at the ice fiord. We decide, in fact, to depart from Sisimiut this afternoon. There is a good harbor forty kilometers to the north at the site of a deserted Inuit village at Ukivik Island—and this, we agree, will be tonight's destination.

*July 26, Tuesday night. Ukivik Island (Sydbay), Greenland.*

The sun at midnight lights the sky with a fiery orange glow. The shadows in the mountains underneath are purple and ultramarine. The visibility is nearly unlimited and the textures of the land are undistorted, as if there were no atmosphere in this place to filter the light.

Tweny kilometers to the east, the flanks of Mount Isortup appear so close that you could reach out and touch them. Fifteen kilometers to the south, a series of peaks on the Natarnivinqup range remain in direct sunlight and burn with the same orange glow. At their base, a thousand meters below, the surface of Nordre Isortoq Fiord mirrors the color so that it, too, seems to be ablaze in the reflected light.

Pete and I stand at the crest of a meadow on Ukivik Island trying to comprehend this landscape. We have come ashore for a few hours after dinner to wander among the remains of the abandoned village here and to investigate a series of grassy mounds scattered about the meadow. These mounds turn out to be the collapsed roofs of traditional Inuit sod houses—the only visible remains of a village that had once been the administrative center and the most important settlement of the region.

We stand facing east, looking toward the mountains. Behind us is the island. Directly in front is the small protected harbor where *Brendan* lies at anchor. The harbor is flanked by a series of skerries and rock ledges. Beyond these are the mouths of several large fiords, including the massive Nordre Isortoq, one of the longest in western Greenland.

In spite of the tiny form of the anchored sailboat before us (or maybe because of it), this place has a strange, unsettling quality to it—as if we simply do not belong here. The physical size of the land is overwhelming. The signs of human passing are faint and quickly covered. There is no enduring human imprint—maybe that's it—or at least none that would indicate that our kind is in any way important to this place. The land exists in a time scale and answers to a purpose that is entirely independent of us—neither benevolent nor malevolent, neither kind nor unkind, but simply indifferent.

Pete and I stand in silence, unable, at least for the moment, to articulate all the feelings that this place evokes. I find myself thinking about time—about the tremendous geologic age of the objects we are looking at. The written record of our species' collective memory, what we call our "history," spans a period of something like ten thousand years. The ice trapped in the glacial bowl behind these mountains spans a period of twenty-five times our history (two hundred fifty thousand years). And the mountains—these span a period of ten thousand times the age of the ice (two and a half billion years) and more. Some of these mountains, in fact, are among the oldest physical objects on the planet.

For some reason, the idea of geologic time seems to have

been resisted by even the most observant of our forebears until very recent times. Why? Maybe because of our distinctly human need to feel that we are at the center of things—that we are cosmically important. This, in combination with our ancient myths of creation, has kept us from noticing the obvious. Yet the truth of geologic time is written everywhere in the Earth—we have simply chosen for thousands of years to ignore the handwriting.

I have a strong feeling as I stand gazing out at these mountains—one that has been growing for the last couple of weeks as we have traveled along this coast and that comes with particular clarity whenever we are confronted by a landscape like this. I feel diminished—unimportant—not unlike the way a sailor feels when he's been traveling for many days on the ocean. Only here the feeling is magnified by the experience not only of limitless space in two horizontal dimensions but also of the massive, superhuman scale of objects in all three dimensions.

Human beings who live their lives in such a landscape must have a different mental image of themselves, I think, than those from gentler, flatter, warmer places. Here the Earth is clearly master. Here there is no illusion of being in control. No amount of self-flattery can mask the fact that under the shadow of these mountains, we are the ones who are visitors—perhaps only temporary visitors at that.

I gaze across the meadow at Pete, then out at the sailboat, framed against the crimson light. Both seem so frail out here tonight, so laughably small. A gust of wind swirls up from the harbor, rattling the grass at my feet. I turn toward the beach where we've left the rubber dinghy, calling for Pete, and I shudder with the sudden cold.

*LOG: Wind: light east. Barometer: 29.10 steady. Weather: overcast, haze. Position: Disko Bay, 30 miles WSW of Illulisat (Jacobshavn), Greenland.*

There are many names for the ice fiord in Disko Bay. On the Danish sailing charts the fiord itself is always identified with the place name of the nearby village: Jacobshavns Isfjord. On the geological survey maps it is sometimes coupled with the name of the ice wall at its mouth: Jacobshavns Isfjeld. In the scientific literature it is named for the glacier that feeds it: Jacobshavns Isbrae. In English we speak of it as the Disko ice fiord or (using the Greenlandic place name) the ice field at Illulisat. But the most expressive name may be the one that the local Inuit residents have traditionally used to describe this place: they call it Kangia, the ice river.

Located on the eastern side of Disko Bay at about sixty-nine degrees north latitude, Kangia is literally a river of moving ice. It begins at the edge of the central Greenland ice sheet, twenty-five kilometers inland, and flows down a deep fiord to the sea. The fiord is between five and eight kilometers wide for most of its length, opening to ten kilometers at its mouth. The rate of movement of the ice as it slips down from the central glacier and calves into the fiord has been estimated at up to twenty meters a day. This rate translates into a staggeringly large total annual volume of some forty-two cubic kilometers—suggesting that this single source may be responsible for fifteen to twenty-five percent of annual iceberg production for the entire west coast of Greenland.

The book of sailing directions that describes the Greenland coast to mariners warns of several dangers to vessels transiting this area. First comes a simple warning about the size and frequency of icebergs to be encountered as one enters and crosses Disko Bay. Then comes a warning about periodic surges in the ice at the mouth of the fiord that can result in sudden, explosive outbreaks of icebergs and iceberg detritus and sometimes dangerous tidal surges.

These outbreaks are caused by a combination of local bottom contours and periodic astronomical events. A shallow sill extends across the ten-kilometer mouth of the fiord—shallow enough that it will not allow the deepest ice to float across. Once several large bergs run aground on this sill, the ice behind them begins to jam up. Twenty-five kilometers upstream the glacier continues to calve—twenty meters per day, every day—and the ice in the fiord mounts, rafts, presses together with increasing force.

The ten-kilometer wall of ice at the mouth becomes increasingly unstable as the pressure builds. Finally, at intervals of about two weeks, the sun and moon align in such a way as to create unusually large "spring" tides. The increases in water level that accompany these tides enable the grounded icebergs to move—sometimes to break explosively into smaller pieces—so that the ice behind them surges forward, spilling hundreds of icebergs across the sill and into the bay beyond. Occasionally these outbreaks are accompanied by sudden tidal waves that can be dangerous to small craft, especially those lying in shallow water near the shore or in crowded anchorages such as the harbor at Illulisat.

———

This morning at four o'clock as *Brendan* proceeds east into Disko Bay, all three members of the deck watch search the horizon, straining for a glimpse of what is ahead. Blue stands at the helm. Mike and Pete watch at the bows for ice. All any of them knows of Kangia at this point is what they have read in the sailing directions and what they have seen on the Danish charts. They know nothing of its actual configuration—nothing of any recent breakouts into the bay.

The instruction I've left with this watch is that someone is to wake me as soon as the ice appears. Finally at five-thirty Pete comes below. He steps into my cabin and shakes me by the shoulders. "There's ice out there now, skipper," he says, "odd-looking ice—lots of it."

I join him on deck several minutes later, and I look for myself at what he's been trying to describe. It is ice all right—a virtual palisade of ice extending halfway across the horizon to the east and showing no land on either side. It glows with an eerie white light, and it is surrounded by a smoky mist that lies along its top and flanks and seems to dissolve into the surface of the sea.

Without other objects around it, the ice presents no clear indication of scale. Are we five kilometers away from something large? Ten kilometers away from something very large? Fifteen kilometers away from something gargantuan? I ask Pete to go below and get a radar fix on the objects we are looking at. Several minutes later he returns with an answer. "There are echoes scattered all over the place—must have been a breakout of some kind. The ones dead ahead are mostly in a line—about twenty-five kilometers away."

It is past eight o'clock when Mike ascends the mast and climbs into the ice barrel. Ahead is a maze of broken iceberg pieces that stream in wind and current rows, like sea ice, blocking the way into Illulisat harbor. To the right is a series of icebergs, the largest of them probably a hundred meters high and half a kilometer long, with narrow, steep-sided channels opening between them. Beyond these is the ice wall itself—ten kilometers of mist-shrouded cliffs carved and broken and blasted into unearthly shapes that tower above *Brendan*'s rigging and obscure all the land beyond.

Pete was right—there has been a recent breakout from the mouth of the fiord. At first there seems no way through the ice to the harbor mouth. Then Mike spots a lead of open water, and another beyond it, and another beyond that. Slowly, he directs the sailboat in a series of looping switch-backs through the ice until at last she has only one more obstacle—a pair of iceberg pieces grounded midchannel in the narrowest section of the harbor mouth.

There is no way around—the ice is too close to the shore. But there is an opening between the two pieces that might be wide enough for *Brendan* to pass—and this is the way Mike now points. The boat eases ahead, skirting the ice with inches to spare, and in another few seconds she is safely inside and we are circling the anchorage basin, looking for a place on the busy commercial wharf to tie alongside.

*July 28, Thursday. Illulisat, Greenland.*

Two American sailboats lie moored in Illulisat harbor this morning. *Bowdoin* lies against a tall steel bulkhead at the Atlantic quay in the outer harbor, while *Brendan* lies in the center of the inner basin, rafted onto a pair of derelict fishing trawlers. The harbor itself is formed of a steep-sided bowl of rock, perfectly protected, with the buildings of the town circling above it. The sky beyond the buildings is deep blue and there is not a ripple of wind on the water—a perfect Arctic summer morning.

Our plan is to spend several days here, climbing the hills behind the town to an overlook above the ice fiord, and then sailing out to the ice wall with *Bowdoin* to witness the ice close up. While we are in Illulisat, only a few miles from the central ice sheet and practically next door to the most active glacial river in Greenland, I am also going to take time to pursue the next phase of the ice riddle.

Conveyor speedup, Conveyor shutdown, unknown quantum states that might exist somewhere in between: it seemed the more I learned about Broecker's Great Conveyor Belt hypothesis, the more questions arose in my mind and the fewer of them Broecker himself seemed able to answer. Could the cold spot anomaly and the increasing sea ice cover between western Greenland and eastern Canada be current-induced? If so, could such changes be the result of a slow-down in the operation of the Conveyor system? Could they even be part of a mechanism that was *causing* such a slow-down?

I knew, even as I started asking these questions, that Broecker was not going to be the one to provide the answers. As his colleague Fairbanks had observed, Wally Broecker is a global thinker. Part of what this means is that he has left much of the development of his ideas to other scientists. In the end, if I wanted to answer specific questions about LDB ice behaviors, I knew I was going to have to look for researchers who had been working on more particular aspects of the Conveyor hypothesis.

With the help of Claire Parkinson, this looking eventually led into the distant past by way of an American geologist named Richard Alley. Like Wally Broecker, Richard Alley is a university professor and something of a global thinker. Among his numerous scientific endeavors, he has been collaborating during the past several summers with a team of American scientists who have drilled a deep core into the central ice sheet at the Greenland Ice Sheet Project (GISP) II site, about four hundred kilometers east of here. The core reaches down through the ice more than three thousand meters to the bedrock at the base of the glacier, and it reveals a series of climatic indicators that have been accumulating in layer upon layer of snow and ice for more than two hundred fifty thousand years.

Perhaps the easiest way to understand the importance of Alley's work is to begin at the end: with an idea that he has recently started writing about but that, as of this summer, he hasn't yet had time to publish. The idea is simple. It is also somewhat daring scientifically, as it runs counter to the conventional wisdom and is derived from evidence that emerges from the ice core record.

*The Arctic and sub-Arctic,* Alley proposes, *and in particular the area around the North Atlantic basin, controls global climate change.*

Such a proposition has a Broeckeresque quality of global thinking about it—and indeed it owes much of its genesis to the influence of Broecker and to the Great Conveyor Belt hypothesis. It is, in fact, an attempt to synthesize a number of apparently disparate pieces of a climatological puzzle that have been emerging from various geological and chemical records all over the Earth, many of them only recently discovered.

"In a community . . . dazzled by the size and energy-richness of the tropics," Alley writes, "it approaches heresy to suggest that the tropics are a slave to puny regions in the Arctic and sub-Arctic. The paleo-record, however, argues that this indeed is the case—the 'accident' of highly variable deep water formation in the North Atlantic causes this region to control global climate on the timescales of its variation."

So here it is—another far-ranging proposition about the LDB area that means the ante in this investigation may have to be raised once again. I'd started by looking for a mechanism to explain an apparently regional anomaly in the sea ice cover of one section of the Arctic. Then, with the aid of Broecker's Great Conveyor Belt hypothesis, I'd learned that events here might be a function of changes taking place in a worldwide system of oceanic circulation. And now, with Richard Alley, I was being asked to entertain the further proposition that this area, along with adjacent areas around the northern Atlantic, does not merely respond to climate changes arising from other geographic regions but in fact is the *global climate control center* where such changes originate and from which they are broadcast to the rest of the planet.

Richard Alley's scientific involvement with questions about ice and climate began in the early 1980s, at a time when the importance of the polar regions was becoming more evident to scientists and when new data about the ice were becoming increasingly available. In the summer of 1981 the drilling of the Dye-3 ice core was just being completed in southern Greenland. That same summer Richard Alley had just finished his first year of graduate study in geology at Ohio State University. Although these events appeared unrelated at the time, Alley had in fact already started working on a problem that was soon to relate directly to Dye-3 and to all future ice coring projects—a problem having to do with the structural properties of glacial ice.

Under the tutelage of glaciologist Ian Whillans, Alley had become fascinated with the transformation that takes place when layers of snow are buried deeper and deeper in a glacier until they are finally compressed into ice. There was a story hidden in the polar ice sheets about how the Earth's climate worked, and Alley knew that if he could learn to read the physical changes that happened during the snow-to-ice transformation, he could add significantly to the knowledge of those who were trying to decipher this story.

Alley's early attempts at solving the snow-to-ice problem resulted, in part, in a study he did with Whillans at Ohio State. The question persisted, however, and eventually it followed him to the University of Wisconsin, where it evolved into the subject of his doctoral thesis. Yet even here, Alley discovered, there were special problems for a young graduate student trying to understand polar ice.

"My first Ph.D. thesis melted!" Alley exclaims in his typically ebullient manner. "I'd spent the winter of 1984–85 in Antarctica, drilling a core for the study. I came home by airplane, and the ice core sections followed in a refrigerated ship. When they finally arrived in Madison in a freezer truck in the spring of 1985, we carried them into the lab, eager to take a look at what we'd got. Only we didn't have anything. During transit they'd all turned into water, then solidified again into meaningless slabs of refrozen ice!"

A second trip to Antarctica during the winter of 1985–86 and a second set of glacial cores finally resulted in both a completed thesis and a solution to the snow-to-ice question. "We figured out how the physical changes happened during glacial formation," Alley says, "so that we learned how to read the ice cores. We learned how to tell the difference between summer and winter, for instance. The ice for each season has a different color, a different texture. This means we could just start counting back, year by year. In the central Greenland ice cores there is an excellent annual record back twelve thousand years—and a very good annual record back forty thousand years—the only record ever found that goes back that far on an annual basis."

Once he'd completed his Ph.D. thesis, Alley began what has evolved into a notable teaching career at Penn State University. He also began working in earnest on several pioneering ice study projects, first in Antarctica and then in Greenland. The most important of the Antarctic studies has been an investigation of internal dynamics on the West Antarctic ice sheet—an attempt to determine how massive ice sheets migrate across the continental bedrock and why they may occasionally surge at much faster, potentially catastrophic rates. This study began in 1986 as a collaborative effort involving a number of scientists and continues into the present.

Several years after the West Antarctic project began, Alley was invited to join another prestigious team of researchers

who were about to begin work on the GISP II deep core
drill project in central Greenland. Alley was selected for this
role because of his earlier work on the analyses of the physical properties of glacial ice. His job for the GISP II project
was to perform similar analyses—counting layers, establish-
ing timetables, looking at physical indicators of annual snowfall amounts.

"This business of snowfalls was particularly meaningful,"
Alley recalls, "as it was the first time anybody had been able
to assemble a long record of precipitation going back into
geologic time. We'd had indicators of temperature, $CO_2$,
methane, all sorts of other chemical properties. Now we
were able to add precipitation to this list—and we started
observing some pretty dramatic things.

"We started seeing snowfall amounts that increased and
decreased very rapidly during sudden warming and cooling
events, for instance. This raised a lot of new questions and
led to lots of inferential stuff about how the ocean/atmo-
sphere system reorganizes itself into radically different modes
in response to particular kinds of forcings."

The basis for Alley's proposition that the Arctic regions
around the northern Atlantic control global climate change
is derived from two sets of information. First are the records
from the Greenland ice cores themselves, especially the two
most recent, the European GRIP and the American GISP II.
And second are various other data sets from around the Earth
indicating that climate events revealed in the ice core record
were felt at about the same times and at similar or smaller
magnitudes all over the planet.

The ice core records are complex, Alley explains, mainly
because the sequence of climate changes that they reveal are
also complex. One set of changes that appears in the record
is related to low-frequency astronomical cycles—the Milankovitch cycles—that occur as the Earth spins on its axis

and orbits around the sun, and these modulate with very large periodicities of between twenty-three thousand and one hundred thousand years. A second set of changes is related to medium-frequency cycles, called Bond-Heinrich cycles, that seem to be driven by internal ice sheet dynamics and that modulate with periodicities of between five thousand and ten thousand years. And a third set of changes is related to high-frequency cycles, called Dansgaard-Oeschger events, whose causes are not yet well understood and that modulate with an abrupt, "square-wave" character with periodicities on the order of one thousand years.

All three types of climate events interact in the ice core record, so that a graph of the result can sometimes look like the wave diagram of a radio with three stations playing at once. For each event, however, the bottom line remains the same. The event occurs first in the Greenland record. Then its analogue occurs, within dating accuracy, in records from other regions across the Earth: pollen records from the sediments of Lake Tulane in Florida; glacial records from the southern Alps of New Zealand or the glaciers of southern Chile; ocean bottom sediments from the Santa Barbara Channel off southern California; deep core records from the Vostok ice core project in Antarctica.

Many indicators thus point to the likelihood of Alley's proposition, although as Alley himself points out, the evidence is far from complete. "No coupled climate model can yet produce all of the observed signals from the observed [Arctic-based] forcings," he writes. "And the incompleteness of paleoclimatic records is troubling."

More records from more sources around the Earth will be needed to confirm the predominance of the Arctic climate events, Alley admits. And other, non-Arctic mechanisms will have to be investigated and modeled. But for the moment, at least, the available evidence points strongly to the North

Atlantic basin as the leading candidate for "global climate control center" and to the Great Conveyor Belt as the most likely mechanism for communicating the Arctic changes out to the rest of the Earth.

As a general statement about the climate dynamics of the area I was investigating, Alley's proposition seemed fairly easy to understand—especially as I was already familiar with Broecker's Great Conveyor Belt hypothesis. But as I became more curious about specific aspects of this proposition, I realized that I was going to have to take a closer look at some of the ancient climate cycles that Alley was dealing with, in terms of both how they functioned in the Arctic and how their effects were broadcast to other regions.

Of the three major types of climate cycles described in the glacial record (Milankovitch cycles, Bond-Heinrich cycles, and Dansgaard-Oeschger events), the ones that were the most dramatic and had the strongest ice core signatures were the Bond-Heinrich cycles. Their discovery, I learned, was also a fascinating story in itself. Beginning several years before they were identified in the GRIP core, they were the subject of a scientific mystery that baffled researchers for half a decade and that both Broecker and Alley had important parts in helping to solve.

The story of these Bond-Heinrich cycles began in the summer of 1988 aboard an oceanographic research vessel off the western coast of France. A young researcher named Hartmut Heinrich, working out of the German Hydrographic Institute in Hamburg, had been drilling bottom sediment cores in the Atlantic Ocean that summer at a depth of some four thousand meters when he made a curious discovery. Buried at regular intervals in the ooze were layers of tiny stones. Heinrich drilled thirteen cores during the course of his investigations, and the same stony layers—usually six in number—occurred in nearly all of them.

Heinrich knew that the stones must have been transported to this area from somewhere else. His candidate for the source (based on the stones' composition of limestone and dolomite) was the region around the Hudson Bay in eastern Canada. His candidate for the vehicle was an iceberg—or more probably a number of icebergs—that had scraped up rocky debris from the continental bedrock before they had broken off from the ice sheet and then had deposited the debris as "dropstones" when they melted.

There was nothing particularly remarkable about the idea that during an ice age stray icebergs could sometimes drift this far east. What *was* remarkable was that during the next few years the same sequence of stony layers with the same geologic timetable and the same composition of limestone and dolomite was found at a dozen other sites spanning the entire Atlantic from Labrador to Europe. It soon became evident that scientists weren't looking just at isolated drop-stones from a few stray icebergs but at oceanwide layers of debris from armadas of icebergs that had invaded the Atlantic in huge numbers.

Other researchers began to take notice of these "Heinrich events," as they came to be called. Among them were Wally Broecker and a group of colleagues from Lamont-Doherty, including geophysicist Gerard Bond. Broecker and Bond and group decided to take a look at the rest of the contents of the stony sediment layers that Heinrich had identified. What they found was that every time the stony debris increased in the sediments, the percentage of shells (or "tests") from the foraminifera plankton (a common organism that lives near the surface of the deep ocean) decreased, often dramatically. Broecker and Bond inferred from these "foraminifera barren zones" that the surface waters of the northern Atlantic had

been intensely cold during these interludes. They then speculated that the absence of the foraminifera tests in the bottom sediments may have been due to a blockage of light at
the ocean surface caused by the sudden inundation of the
Atlantic by huge numbers of icebergs.

In 1991, when this study was reported, neither the GRIP
core nor the GISP II core had yet been completed, and there
were no existing Greenland ice core records that showed any
pronounced features corresponding to the Heinrich events.
All this was soon to change, however. The European (GRIP)
team was a year ahead of the American (GISP II) team. The
GRIP drilling was completed in 1992 and the analysis of the
core was completed a year later. The resolution of the temperature record in this core was far superior to anything
scientists had seen before, and the Heinrich events were
clearly visible.

It was Broecker's colleague Gerard Bond who actually
identified the cycles that contained the Heinrich events (thus
the name "Bond-Heinrich cycles"). Through a correlation
of the deep-sea sediment record containing the Heinrich
layers with the new GRIP ice core record, Bond showed
that the shorter-phased (Dansgaard-Oeschger) cooling cycles
were actually bundled into sequences of progressively colder
episodes. Each sequence ended with a Heinrich-type layer of
dropstones, followed by a sudden period of intense warming
and the beginning of a new Bond-Heinrich cycle.

It seemed obvious, in light of these correlations, that the
hypothesized iceberg armadas were indeed the mechanism
that had created the Heinrich layers—also that these sudden
influxes of cold freshwater to the surface layer of the North
Atlantic had somehow altered the operation of the Great
Conveyor Belt, forcing it to flip into another of its quantum
steady states and leading to the abrupt rise in temperatures
over Greenland. Now the question was what had caused the
armadas? Why, at a particular point in each Bond-Heinrich

cycle, had the huge Laurentide ice sheet over central Canada suddenly surged down through the Hudson Strait to flood the Atlantic Ocean with icebergs?

An answer to this final part of the mystery came during a meeting of the American Geophysical Union in December 1993, at the hands of a pair of researchers: Doug MacAyeal and Richard Alley. For the past several years MacAyeal (incorporating some of Alley's ideas) and later MacAyeal and Alley together had been studying the dynamics of ice sheet surging on the West Antarctic ice sheet. When they began reading about Heinrich events, they realized almost immediately that there was an analogy between the way the ice was moving today in Antarctica and the way it must have moved during the last ice age over what is now Hudson Bay.

During their studies on the West Antarctic ice sheet, they had discovered a curious fact. When the bedrock underneath a glacier is compressed under a thick, insulating blanket of ice, heat emanating from the Earth's inner layers is trapped until the bedrock reaches its melting point, pressurized ice is turned into water, and soft sediments are transformed into a slurry that is roughly the consistency of toothpaste. When this happens, the toothpaste-slurry provides lubrication that enables the overlying layer of ice to move quickly—much the way the ice river flows down from the glacier at Kangia—at speeds on the order of tens of meters per day.

In much of the continental area of eastern Canada that supported the Laurentide ice sheet, the bedrock is hard and crystalline. But in the Hudson Bay area the layers of limestone and dolomite are softer—soft enough, MacAyeal and Alley speculated, that when the ice sheet above them grew to a thickness of about two thousand meters, the heat at the bottom warmed the bedrock and melted the ice, allowing the water and soft sediments to form into the toothpaste-slurry and the ice sheet to begin sliding toward the sea.

In this particular geographical region there was a natural

gateway for the ice to move: a hundred-fifty-kilometer-wide opening between the coastal mountains of Labrador and Baffin Island called the Hudson Strait. As the ice flowed through this gateway, it pushed out into the Labrador Sea and began calving into hundreds of pieces, creating precisely the sorts of iceberg armadas that the Heinrich layers foretold.

As to the timing of the Heinrich events, MacAyeal also had an explanation that seemed logical. His calculation of snowfall amounts for eastern Canada during the ice ages indicated that it would require on the order of seven thousand years for the glacier in this region to grow to the height it needed to trigger such an episode—almost the identical timetable that had been revealed for the actual events in the ice core record.

For Richard Alley there was one more question that needed to be answered about the Bond–Heinrich cycles. Was there evidence, beyond the ice core record, to show that these massive iceberg surges somehow got broadcast around the world by the Conveyor Belt to become global climate events?

The answer, Alley found, was yes. A careful review of ancient climate signals emerging from locations around the globe indicated that indeed, there *was* such evidence—albeit in widely scattered places. The pollen records from the bottom sediments of Florida's Lake Tulane, for instance, indicated a clear change between warmer, wetter climates and drier, windier climates on the timescale of the Bond–Heinrich cycles. Even more surprising, data from southern Chile during each of the four most recent Bond–Heinrich cycles suggested that there had been surges in the maximum ice extents of Andean mountain glaciers that correlated well with the Bond–Heinrich timetables.

If these correlations hold up to scrutiny—and if others are found in other parts of the world to supplement them—

then, Alley reasons, the conclusion will be inevitable: the northern Atlantic basin will indeed be recognized as global climate control center, and events that happen here will be understood as the ones that ultimately drive the rest of the planetary system.

*July 29, Friday. Illulisat, Greenland.*

The village of Illulisat is located on a peninsula about three kilometers wide. To the north is the harbor basin where *Brendan* lies moored. To the south, a kilometer and a half beyond the town and just over the crest of a series of small hills, is Kangia, the ice river.

Soon after breakfast my crew begins to move ashore, first to wander through the village, then to find their way out to the ice fiord and the rugged countryside beyond. I am last to leave the boat. I see no sign of the others as I step across a pier and climb a wooden staircase at the head of the harbor. I walk through the town along a road that ascends into the hills, then I follow a footpath that winds around deep fissures in the rock toward the farthest summit.

The terrain here is covered with black lichen. The sky is dark blue. At first I see only these two colors meeting at the crest of the hill ahead. Then, as I climb the last few feet toward the summit, a thin line of white appears between them, growing wider with every step. In another few seconds it becomes a blinding, utterly improbable sea of white—an object so bright that at first there is no detail, only a painful blur.

I stand at the summit and allow my eyes to adapt, slowly, to the vast sweep of ice before me. The silence is nearly complete, interrupted only by an occasional creak or groan somewhere far in the distance. The texture of the ice surface is wild; its scale is immense. The feeling it gives is not of something solid, however, but of just the opposite: something tentative and highly volatile—something that is waiting to crack, raft, slip-fault, explode, transform itself into a thousand new forms.

Part of the effect that this place has on me has to do with its sheer size. Another part has to do with my understanding, however incomplete, of the massive geophysical forces at work here. For several minutes I stand and gaze at the ice, my mind spinning with a jumble of disconnected thoughts. Then I find myself thinking again about Richard Alley, and in particular about a question that I'd started asking soon after I'd first encountered his ideas—a question that seems particularly relevant to this place.

From the moment I'd started thinking about Alley's global climate control center, I'd understood how it might work during an ice age—at a time when huge glacial ice sheets covered much of the Northern Hemisphere and when armadas of icebergs (or other massive ice events) could shut down the Great Conveyor Belt.

But these were, by definition, ice age occurrences. What about an occurrence that might take place during a warm interglacial period? Was there *any* evidence that a catastrophic climate shift could happen now—at a time when there were no other major glaciers except this one anywhere around the northern Atlantic?

Until recently, the preponderance of scientific opinion has

supported an almost unanimous answer to this question. Beginning with the analysis of the Camp Century drill core in southern Greenland in the late 1960s, scientists have found increasing evidence that ice age climates have been highly volatile—shifting from cold to warm and back to cold—and that such shifts have often been quite sudden. On the other hand, records from our own ten-thousand-year warm period (the so-called Holocene period) do not show any such tendency. In fact, since the retreat of the last great continental ice sheets and the beginning of human history, virtually all the evidence that scientists have assembled indicates that the Earth's climate has been amazingly stable. The pollen records, the lake and ocean sediments, the tree ring thicknesses, the ice core data all point to the conclusion that there haven't *been* any significant climate events since the end of the last ice age—at least no major ones.

In light of such findings, most scientific observers have concluded that interglacial climate has alway been comparatively stable and that large, sudden climate shifts, when they have occurred, have taken place only during ice ages. But there is a new source of evidence that may challenge this assumption—a set of findings from the European (GRIP) ice core project, published in the summer of 1993, suggesting that a hundred twenty thousand years ago during the last warm period (the so-called Eemian period), the climate may have been just as volatile and the shifts may have been just as sudden as they have been during ice age climates.

According to the GRIP team's analysis, the Eemian warm period behaved as a three-state climate, with one state about ten degrees Celsius warmer and one state ten degrees colder than our own and with large, sudden shifts between them, sometimes taking place within a matter of decades. Our own Holocene climate, in contrast, has displayed only one state. We've had some minor variations since the last ice age—

global warmings and coolings of perhaps two or three degrees Celsius. The latest was a five-hundred-year cold snap
called the Little Ice Age that ended about the middle of the
last century. But the Little Ice Age was a minor blip compared with the shifts that researchers seem to have identified
in the GRIP-Eemian. Those shifts were really large—three
or four times as large as the shift in the Little Ice Age—and,
according to the GRIP analysis, they were also quite sudden.

The potential importance of the GRIP team's findings is
clear. If the last interglacial climate could indeed flip back
and forth between radically different stable states, then our
own climate might be able to do the same. And if this is the
case, then a climate change that was assumed to require hundreds or even thousands of years might happen in a matter of
decades, causing major disruptions in the lives of thousands
of living species and millions—perhaps billions—of human
beings.

Alley, one should note, is not ready yet to comment on
the GRIP team's findings. He has been a member of the
American team that has been drilling at the GISP II site. The
two sites are located about twenty-five kilometers apart at
the summit of the Greenland ice sheet, close enough that
they'll eventually be used to verify one another's findings.
But because of early equipment problems experienced by the
Americans, the European team was able to get a year's head
start on the drilling—and this means the analysis of their data
is also a year ahead. Alley and his colleagues probably won't
be ready with their analysis until late this summer.

Meanwhile, the ice core record for the GRIP-Eemian
stands as a warning—albeit a tentative one—to a species that
has been tampering with the chemistry of its planet. Many
questions remain. The analysis of the GISP II data may answer some of them. Analysis and scrutiny of other deep ice
cores to be drilled in the future will undoubtedly answer

others. In the end, however, the experiment that we are conducting upon our planet will provide its own answer—one that will not be open to interpretation or debate and that may affect every region and every living creature on the Earth.

Maneuvering in close quarters: Gotthabsfiord
PHOTO: MYRON ARMS

Mike Browne (Blue)
PHOTO: MICHAEL AUTH

Amanda Lake
PHOTO: MYRON ARMS

*Brendan's Isle*
PHOTO: MYRON ARMS

Pilot whales in the Davis Strait
PHOTO: MYRON ARMS

Pete and Amanda on deck
PHOTO: MYRON ARMS

View from the hills above Kangia

Schooner *Bowdoin* framed against an iceberg

Skipper Mike Arms
PHOTO: AMANDA LAKE

Sailing in ice: Labrador Coast
PHOTO: MYRON ARMS

Growlers—*Brendan's Isle* in foreground
PHOTO: MYRON ARMS

Light winds: Disko Bay
PHOTO: MICHAEL A. BROWNE

Arctic light: Disko Bay
PHOTO: MYRON ARMS

Maneuvering close to the ice
PHOTO: MICHAEL AUTH

Floating ice mountain
PHOTO: MICHAEL AUTH

Ice shapes: Labrador Coast
PHOTO: MYRON ARMS

*July 30, Saturday. Claushavn Anchorage, Greenland.*

The lower limb of sun intersects the rim of the sea this evening just as the ship's clock chimes midnight. All my shipmates are sleeping below, and I'm alone on deck standing the first shift at anchor watch. *Brendan* lies in an open roadstead, anchored behind a rocky outcrop about five kilometers from the southern edge of the ice wall. The wind is calm. An oily swell works its way past the outcrop and rolls *Brendan* side to side. An iceberg grounds itself on a ledge a hundred meters ahead of the boat, and another moves slowly past the stern in the current.

There is no protection in this anchorage—only open water for a hundred fifty kilometers across Disko Bay and Davis Strait to the edge of the Baffin ice pack. But there is a village of several dozen traditional wood buildings on the hill to the east and an unobstructed view of the ice wall beyond. To the north of the wall, eighty kilometers across the bay, is Disko Island itself—as clear this evening as if it were only a few kilometers away.

Now, as the ship's clock finishes the last stroke of midnight, I sit alone under the spray dodger at the top of the companionway and gaze at a red circle of sun—a midnight sun—as it hovers just above the surface of the water and lights the scene before me with an eerie glow. I know that in literal fact there will be no twenty-four hours of sunlight here tonight. It is the end of July—five weeks and more past the summer solstice—and we've not quite caught up with the circle of continuous light as it recedes northward.

Yet there is an odd phenomenon that happens with the sun at these high latitudes—a bending of the light rays when they are within a few degrees of the horizon—so that even when the real sun is set, the refracted image of the sun still appears to be above the horizon. Thus, according to the evidence of my own eyes—and in spite of strict astronomical rules—*Brendan* has sailed to the midnight sun tonight and I stand witness. I feel the rush of adrenaline that comes with twenty-four hours of daylight—a surge of almost boundless energy in which sleep seems irrelevant. I feel awed, as so often on this voyage, at the magnificence and impersonality of this vast system of water and rock and biomass as it turns its continual dance around the sun. And I feel sad—because I realize that this moment also marks *Brendan*'s farthest progress north.

Farthest north—a moment to be reckoned with in any Arctic voyage. Yet this one is different, for this moment, powerful as it is, is but another step in a process whose shape is not a straight line but a circle. Like the voyages of John Davis, ours has never been conceived as a dash toward the Pole. Instead, the northing we've achieved has been accomplished on the way to somewhere else.

For the next week we will avoid the ice to our west by doubling back along the Greenland coast—time for a continued encounter with this landscape, and time, as well, to pursue the ice riddle back into our own era and to explore questions about present-day mechanisms that may be affecting the operation of the Great Conveyor Belt. Then, once we're clear of the Baffin pack, we'll continue across the Davis Strait toward a coast that has always been my own personal goal for this voyage and that promises to be the most dramatic landscape we will visit all summer—the Tourngat region of northern Labrador.

# VI.

# South

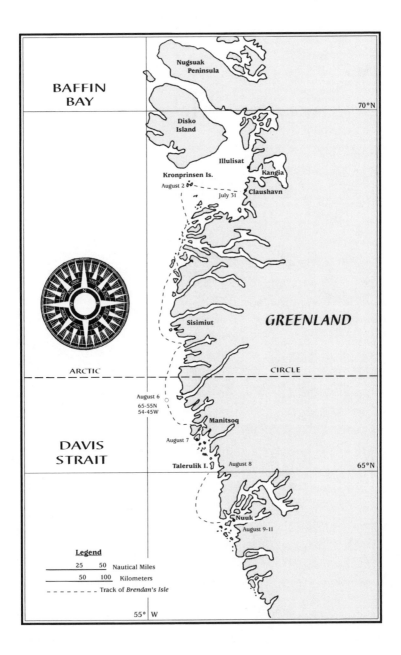

BAFFIN
BAY

70°N

Nugsuak
Peninsula

Disko
Island

Illulisat

Kronprinsen Is.          Kangia

August 2          Claushavn
        July 31

Sisimiut          *GREENLAND*

ARCTIC          CIRCLE

August 6
65-55N
54-45W

Manitsoq

August 7

DAVIS
STRAIT

Talerulik I.          August 8          65°N

Nuuk
        August 9-11

**Legend**

25        50   Nautical Miles

50        100   Kilometers

Track of *Brendan's Isle*

55°   W

*LOG: Wind: south, 25 knots. Barometer: 29.40 steady. Weather: overcast, fog, rain. Position: 10 miles due E of Sondre Stromfiord, en route toward Manitsoq.*

The weather has turned contrary on this second transit along the Greenland coast. Five days ago *Brendan* rode out a gale in a deserted anchorage in the western approaches to Disko Bay. Three days ago another gale raged in an anchorage in the Iserquk Fiord, with katabatic winds tumbling down out of the mountains and buffeting *Brendan,* flipping and nearly sinking the rubber boat. Ever since these two systems have passed there have been rain and fog and more south wind—enough to kick up an uncomfortable chop in the north-flowing current and make the going slow and tedious.

Today we are motorsailing—using the engine in combination with the sails—in an attempt to punch through square-shaped seas and keep the boat moving forward. The drone of the engine, the uncomfortable motion of the boat, the cold air temperature, and the crowded conditions below all combine to make the day seem long, and after many hours of running this way, the passage begins to feel like work.

Everyone is trying to keep a positive attitude, and most are succeeding. This afternoon, however, I am having a problem again with Blue. He mumbles to himself while we're in the cockpit together, and he glowers every time our eyes meet. He hasn't said a civil word to me for the better part of twenty-four hours. He's obviously angry and—I'll admit it—his tactics are beginning to get to me.

What's the matter with this guy, anyway? Am I going to have to go to war with him every time I want to use the engine for the rest of the summer? The lines of battle between us are becoming more and more clearly drawn, and the conflict seems to be escalating. This whole affair is beginning to carry over into other parts of our relationship. It's not good, not good at all.

I know what's bothering him—I've known it all along. And the funny thing is, I *agree* with most of what he's been trying to tell me. Hell, I can give the same sermon he gives. I know all the carbon dioxide statistics by heart. I know that the average internal combustion engine produces two tons of carbon dioxide a year. I know we've got half a billion such engines in the world right now, pouring a billion tons of $CO_2$ into the atmosphere annually. I also know that a billion tons is only one-sixth of the grand total—so that as a species we're producing twelve times our collective body weight in atmospheric carbon every year—nearly *six billion tons.*

The trouble is, I also know a lot of other things—things that implicate both me *and* Blue and make us both part of the problem. I know, for instance, that this boat is made of fiberglass and the sails are made of Dacron—both materials fabricated from hydrocarbons and both eventually due to enter the waste stream that we collectively produce. I know the car Blue drives at home produces the same amount of $CO_2$ as the car I drive at home—and that both produce several times more annually than *Brendan*'s diesel. And I know some of the environmental costs of the new house that

Blue's parents have recently built—"costs," that is, in terms of manufacturing pollutants, nonrenewable resources used, $CO_2$ produced during heating, cooling, electrical use, and so forth.

I'm about ready to join battle with my idealistic young shipmate, to raise the ante and begin pointing my finger back at him. But I don't, for two reasons. I don't, first, because I know I won't change his mind. He is in no mood to accept the notion that we may all be part of the problem—as consumers, as willing participants in a society that has not yet learned how to recognize or respond to the environmental destructiveness of its appetites.

And I don't, second, because I don't really want to change his mind. His idealism represents a hope—that the ground rules can be challenged and that our collective behavior can be changed. If a large enough number of Blue's generation were to feel as he does, and if each person were to challenge my generation (along with his own) to make the necessary adjustments in the choices we make, then maybe—just maybe—we could begin to undo some of the damage that we've done.

Blue is angry at me because he is looking for a hero—somebody he can idealize and emulate—and I'm just not measuring up. Don't get me wrong—I don't enjoy being a bad guy. I'd *like* to measure up. But I've been around too long, I'm afraid. I'm an old man in Blue's eyes, a member of the offending generation, and we are locked in a classic battle between youth and age.

Blue's idealism is seamless; mine is cankered. His admits of no exceptions; mine is full of compromise. He still thinks he can change the world—single-handedly. I no longer think we can do anything—single-handedly.

He misunderstands me—or so I tell myself. I'd like to be the hero he is looking for, but the only important thing right now—the only necessary thing in terms of the objectives of this voyage—is that I remain the skipper of this boat and he remains the crew. I don't like the battle we're in, but I'm as stubborn as he is, and I will not capitulate to his mute demands any more than he will compromise to mine. So I put up with his dirty looks and let him grumble, and I say nothing.

There is still no useful weather forecast for this coast, but once again the evidence of our eyes is enough to tell us that there's another dose of nasty weather brewing. All afternoon as we wind our way through an inner passage, the signs multiply. The barometer begins to fall. The overcast to the south turns increasingly into a steely gray. The wind becomes fitful and squally. Mike calls me to the nav-station at one point to show me a weather-fax map for the Labrador Sea, a thousand kilometers to our south. He points to a depression that has moved across that region out of Hudson Bay this afternoon, following the same pattern that we've noticed on these maps for the past week. If I were a betting man, I say to Mike, I'd put my money on another gale on this coast by sometime tomorrow.

We're about to enter a fiord system known on the charts by its Danish name, Hamborgersund. Fifty kilometers along this channel is the village of Manitsoq, a stopping place that we're familiar with from the passage north. With another bout of weather coming, I suggest that we try for Manitsoq tonight—at least there we'll see a few new faces and maybe have a place to stretch our legs while we're waiting out the next gale.

*August 7, Sunday morning. Manitsoq, Greenland.*

The Danish name for the place we're located this morning is Sukkertoppen—Sugartop. The traditional Inuit name is Manitsoq—the Uneven. A glance upwards at the jagged series of peaks that surround the settlement suggests that, at least as a geographical description, the Inuit name is the more appropriate.

The small harbor basin in front of the settlement is almost completely landlocked. The village itself lies to the east of this basin and comprises bright-colored buildings, many in the traditional Danish style, with steep pitched roofs and board-and-batten siding. In the center of these structures is the fish processing plant, and in front of the plant, jutting into the harbor, is the commercial pier where *Brendan* lies rafted alongside a local search-and-rescue vessel named *Malik*.

The morning begins with a practical mission of some urgency. There was no freshwater hose of appropriate diameter in the harbors to the north for filling *Brendan*'s water tanks—

which means she has not taken on water for nearly two weeks. Rumor has it that there's good water here at the fish plant if we can only find the proper official to ask. This morning after breakfast, Amanda sets out to investigate the plant area and see who she can find, and soon she returns not only with a freshwater hose but also with a cheerful young Dane who introduces himself as Peter Beck, manager of the plant.

Mike and several others attend to the task of filling the water tanks, while I invite Beck to stay for a coffee. I'm eager to learn more about Manitsoq and about the experience of a European who has been living twelve months a year on the Greenland coast.

As we begin to talk, I realize almost immediately that this man is about as far as one can be from the stereotypical European entrepreneur out to plunder and profiteer. There is nothing of the soldier of fortune about this man, nothing of the old arrogance that once typified whole generations of Western colonialists. For one thing, this European is married to an Inuit woman. For another, he speaks fluent Greenlandic (the Inuit language) as well as near-flawless English. He is soft-spoken—almost shy—yet he is obviously well educated. His narrow features, neatly trimmed beard, and wire-rimmed glasses make him look more like an Oxford don than a Greenland fish plant manager.

We begin by discussing the plight of the fishery here and the diminishing tonnages being processed by the plants all over Greenland. "But this is not what we should be talking about on such a pleasant morning," says Beck, clearing his throat and glancing over at me with a vexing smile. "Tell me why you have made such a long voyage to Greenland. Tell me of your purpose here."

I describe something of our two months' journey and of my curiosity about the behavior of the sea ice in this area.

"I've been looking into mechanisms for changing global climate," I say, "asking questions about causes, human-induced and otherwise. Western Greenland seems to be a particularly sensitive area in terms of climate signals—a place where a few answers to these questions may be hiding."

"It's curious that you should be interested in the ice here," says Peter, "because in the four years that I have lived in Manitsoq, there have already been two uncommonly icy winters."

"Are you referring to *sea* ice?" I ask.

"Sea ice—yes. Here around Manitsoq is a part of the ocean where the water is supposed to stay open and ice-free all year. And indeed it does—most of the time. But not in the winter of 1992. Then we had ice in the harbor and all the surrounding bays for February and half of March. Also in 1993—heavy ice again for much of the late winter."

I tell him that I'd been aware of these events, that I'd actually watched them unfold in a sequence of Navy/NOAA ice analyses during both winters. "The heavy ice along the west Greenland coast was quite dramatic on the charts," I say, "especially in 1993. In that year the whole open water indentation in the Davis Strait that's supposed to be kept ice-free by the current was jammed with pack ice during two weeks in March—almost as if the entire West Greenland Current had slowed or stopped for a time."

"I cannot talk about currents," says Peter, "as I do not know of such things. But to have seen Manitsoq harbor filled with pack ice so that not even the coastal freighter could make it through—this was a rare thing. The old people say they have seen this happen only once before, and then only for a single winter in the early part of this century. I have lived in Manitsoq four years and have already seen the ice form here twice. This is a very rare thing—everyone in the village will tell you so."

Peter continues talking, telling about various other aspects of his life. I listen, but with only half an ear, as my mind is stuck on what he was saying about the two anomalous ice years here. I remember something Claire Parkinson once wrote: something to the effect that a change in winter sea ice along the coast of western Greenland might serve as an indicator of regional changes in thermohaline circulation—perhaps even as an early warning signal of climate change on a much larger scale.

I find myself thinking about massive geophysical processes with strange and provocative names: the Great Ocean Conveyor Belt, the North Atlantic Deep Water, the halocline catastrophe, the Great Salinity Anomaly. These are not subjects I will be able to discuss with Peter—even though I suspect he might find them an intriguing framework for explaining the events he has witnessed.

As Peter and I continue talking, Mike shuts off the flow of water to *Brendan*'s tanks, and he and Blue drag the hose across the deck and roll it up on a large metal wheel in a building at the head of the pier. Soon afterward Peter stands and announces that he must return to the plant to prepare for the arrival of several local trawlers. I thank him for his kindness, and I thank him as well for our conversation. "We may need to spend the next day in Manitsoq," I say, "with gale warnings posted again for this afternoon and tonight."

"Stay as long as you like," says Peter. "The pier will not be busy. There are local fish and seal meat for sale in the village. And the Seaman's Hostel is only a few hundred meters down the road for bathing and a hot meal. I believe you will find our village a pleasant place to visit."

*August 7, Sunday late afternoon. Manitsoq, Greenland.*

The wind howls in *Brendan*'s rigging and the rain drives hard against the deckhouse windows. The stove chimney hisses as the rain runs down through the deck fitting. A deep swell finds its way into Manitsoq harbor and surges around the basin, causing the boats to pitch and strain at their moorings.

An hour ago *Malik,* the search-and-rescue vessel that *Brendan* has been moored alongside, was sent out on an emergency call to search for an open boat, overdue and presumed trapped in a nearby bay. *Brendan* was required to cast off and move around to the north side of the fish plant pier, where she's now been resecured in a more protected location. Meanwhile, the tide has dropped more than ten feet during the afternoon, and the sailboat has now practically disappeared at the foot of a weed-covered bulkhead. Until the water returns, there will be no easy means of access or egress and my crew and I will be trapped here—safe, but virtually cut off from contact with people ashore.

I decide to take advantage of this enforced confinement by

continuing with the ice investigation—and in particular by pursuing the question I'd been asking a few days ago about present freshening mechanisms that might lead to a sudden shift in the operation of the Great Conveyor Belt.

Part of the answer to this question will involve the local ice anomaly described to me this morning by Peter Beck. But the most important part of the answer—the part that must come first—will involve a larger and far more dramatic anomaly that scientists have been observing all around the North Atlantic basin for the past several decades, a phenomenon that's been described as "one of the most persistent and extreme variations in global ocean climate yet observed in this century": the Great Salinity Anomaly.

My efforts to learn about the Great Salinity Anomaly (or GSA, as it is referred to in scientific circles) began a few months after I'd encountered Richard Alley and had started thinking about his ideas on the importance of the northern Atlantic basin as global climate control center. As so often in this process, when I'd started wondering where to turn next, I found myself traveling back to Goddard Space Flight Center to talk with Claire Parkinson. Once again I was in need of scientific advice. I knew that she would listen to my questions, and I had a hunch that she might have a few useful suggestions.

"I think it may be time now to start bringing this investigation back into the present," I said to her. "Is there anybody out there who's been paying attention to how global thermohaline circulation may be operating today? Is there anybody monitoring deep convection or talking about present-day mechanisms that might be changing the behavior of the Conveyor?"

Parkinson smiled at what I could only imagine was my scientific naiveté. "Of course," she said. "There are many

such people. In fact, one who's been doing some really interesting work in that area has an office right next door.
Would you like to step around the corner and meet her?"

Several minutes later I found myself standing in the middle
of a workspace similar to the one we had just left, surrounded by computer terminals and shelves of books and
folders, being introduced to another member of NASA's
Oceans and Ice Branch, oceanographer Sirpa Hakkinen.

Parkinson briefly described for her colleague the nature of
my investigation and the purpose of the journey I was planning to make the following summer. Then she left us for a
time so that I could pursue a few of the questions that I'd
come here to ask.

Hakkinen began by telling me about how she had come
to her interest in Arctic sea ice. She was a physicist, she said,
trained in her native Finland before emigrating to this coun-
try to study oceanography at Florida State University. After
completing her Ph.D. thesis—a computer-driven numerical
model of ice-ocean dynamics in a small, idealized setting—
she had turned to the construction of a much more ambi-
tious numerical model of ice-ocean dynamics for the entire
Arctic basin.

Perhaps not surprisingly, her level of diction as she talked
about this work was highly technical. This, combined with
her Scandinavian accent and her enthusiastic, sometimes almost breathless manner of speech, often made it difficult for
me to follow her. She was eager to help, however, and when
she realized that I hadn't understood certain aspects of her
work, she pulled out copies of several of her recent publications and suggested that perhaps I do some reading, then that
we talk again. She also found a copy of a piece of scientific
detective work written about five years earlier by an English
oceanographer named Robert R. Dickson, the original and
most important chronicler of the Great Salinity Anomaly.

"Read this first," she said. "Once you understand what

it's about, you'll find it a lot easier to understand the work I've been involved in here.

"The most recent model run we've done at Goddard spans a period of thirty years—from 1955 to 1985—and describes the dynamic mixing of very large masses of water along the ice-ocean boundary. During this time an important event has taken place in the northern Atlantic—a massive freshening event. Our model imitates the behavior of this event quite successfully and shows one possible explanation for why it took place. But to understand the model, as I say, you must first understand the event itself."

The scientific discovery of the event Hakkinen was talking about, the Great Salinity Anomaly, began with an odd and possibly alarming occurrence first observed in the summer of 1976. In June of that year a Scandinavian hydrographer named H. D. Dooley noticed a strange alteration in the chemical makeup of the ocean in the area between the Shetland and Faroe Islands, several hundred kilometers north of Scotland. The surface and near-surface water all across this area, down to a depth of about five hundred meters, had suddenly grown anomalously fresher.

Dooley and several colleagues tried to imagine what combination of geophysical forces could possibly have triggered such a large change in the ocean's chemistry. Their explanation—which involved a massive eastward shift in the circulation pattern of the entire northern Atlantic basin—has not withstood the test of time. Nevertheless, Dooley has been credited with having alerted the scientific community to a major climate event—the freshening of a huge part of the northern Atlantic Ocean—and to its possible physical and biological consequences.

The next scientific observers to have noticed a massive

pulse of fresh surface water in the northern Atlantic were a pair of researchers named Pollard and Pu. This time the freshwater was observed along the European side of the Atlantic down to the latitude of northern Spain, with its strongest signal occurring in 1977, roughly the same time period as the Dooley observation. Pollard and Pu also tried explaining the anomaly they had observed, although once again they were unable to devise a convincing scenario.

It was several more years before the whole story of the Great Salinity Anomaly was finally pieced together—and it required a clever bit of scientific detective work by researcher Robert Dickson and colleagues to finally assemble all the parts and form them into a coherent whole.

Little by little during the late 1970s and early 1980s, evidence had been mounting from a variety of sources that something highly unusual had been taking place around the North Atlantic basin. In addition to the Dooley measurements and the Pollard and Pu study, there had been reports of a large influx of fresh surface water at numerous other ocean-monitoring locations—ocean weather ships, oceanographic research sites, fishery stations—all the way from the Labrador shelf in the west to the Norwegian Sea in the east. It was left to Dickson and colleagues to piece all these seemingly unrelated reports together—a task he was able to accomplish by means of what he termed an "advective explanation."

First, Dickson noted, there had been a distinct shift in the pattern of winds and currents in the area east of Greenland and north of Iceland during the second half of the 1960s. In response to this shift, the temperature of the ocean surface there had started to fall, the salinities had started to decrease, and a large area of sea ice had been allowed to form, increasing total sea ice cover and inhibiting deep convection across the region for a number of winters.

The sea surface east of Greenland reached a temperature minimum in the winter of 1967 and a salinity minimum in the winter of 1968. Under the heavy ice cover, the formation of Greenland Sea Deep Water (GSDW) either slowed or stopped altogether, and a kind of ice age condition descended upon the region.

According to Dickson's advective hypothesis, the next chapter of this story took place as this huge pulse of fresh surface water moved out of the Greenland Sea by means of the East Greenland and Irminger Currents, swept around Cape Farewell at the southern tip of Greenland, and entered the Labrador Sea/Davis Strait area by means of the West Greenland Current. Records indicate the passage of a massive pulse of freshwater across a section of the Fyllas Bank off western Greenland in about 1970. Somewhat later, a pulse of approximately the same size was observed off the mid-Labrador coast, with its peak passing there about 1972. Estimates of the salt deficit represented by this event run on the order of seventy-two billion tons.

From the Labrador coast, Dickson traced the progress of this massive freshwater pulse around the current system of the entire northern Atlantic—a counterclockwise pattern of surface and near-surface flow that scientists refer to as the "subpolar gyre." After noting its passage across the Grand Banks in the early 1970s, Dickson followed the event eastward across the Atlantic to ocean weather ship *Charlie* in 1974, to the Rockall Channel in late 1975, to a weather station south of Iceland in 1976, to the coast of western Norway in 1977–78, to Spitzbergen Island in 1978–79, and finally back to the Greenland Sea in early 1982.

According to Dickson, there are two simple tests by which to assess the plausibility of his advective hypothesis. First is to determine whether the *magnitude* of the anomaly, as it was observed at various points around the North Atlantic, was

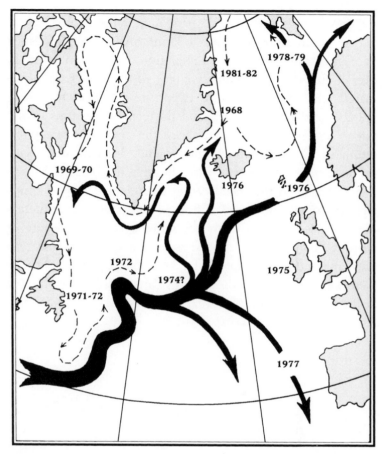

**Dates of the Great Salinity Anomaly superimposed
on a current diagram of the northern
North Atlantic.**

consistent with its proposed character as an advective event.
And second was to determine whether the anomaly's *speed*
was consistent with what we know about the speed of the
ocean surface currents in this region.

First as to magnitude, an estimated salt deficit of about
seventy-two billion tons near the Labrador coast in the early

1970s was followed several years later by an estimated salt deficit of about forty-seven billion tons in the Shetland-Faroes Channel, suggesting a reduction of about one-third in overall size during the intervening four years. Later, estimated salt deficits of about twenty-four billion tons near Spitzbergen Island in 1979 and thirteen billion tons in the Greenland Sea in 1982 completed the circle. As best-guess estimates, Dickson argues, these values are consistent with the idea that the anomaly was advective in character.

Second, as to the speed of the event, the 1968–82 timetable translates into a rate of flow of about three centimeters per second—almost exactly the same as the average rate of current flow for the subpolar gyre. Partly from these two tests, magnitude and speed, Dickson's advective hypothesis continues to stand as the best explanation anyone has yet devised for the fourteen-year saga of the Great Salinity Anomaly.

*August 7, Sunday evening. Manitsoq, Greenland.*

A gust of wind rolls *Brendan* toward the weed-covered side of the fish factory pier, causing a backdraft in the diesel heater. The flame flickers and turns dark yellow. I look up from the jumble of folders and scientific papers that lie on the dinette table in front of me, and I watch a curl of greasy smoke rise from the heater lid. A few seconds later the wind eases back, the smoke disappears, and the heater draws again with a baritone hum that resonates through the boat. As I turn my attention again toward the table where I'd been reading, I realize that my mind is filled with questions.

The Great Salinity Anomaly, it seems, has now been verified as a bona fide climate event—which means that scientists have finally identified a present-day freshening mechanism with the capability, at least, of capping deep convection in the northern Atlantic and slowing the Great Conveyor Belt—possibly even shutting it down completely. With such a mechanism in place, it becomes possible to hypothesize sudden, very large shifts in global climate during a warm interglacial period such as our own—shifts that might occur within a matter of decades and that might affect huge geographic areas and large numbers of living organisms. With the reality of a mechanism such as the GSA staring us in the face, the specter of the GRIP-Eemian becomes more onerous.

But there are still dozens of unanswered questions. Do scientists have any idea yet what caused the GSA? Do they know how it may have influenced the operation of the Great Conveyor Belt or whether it has yet broadcast its effects out into the rest of the climate system? Do they know about its history? Do they know, for instance, whether something like the GSA has ever happened before? Do they know if it could ever happen again? Do they know if it could be happening again *right now?*

I realize that answers to some of these questions have not yet been found and answers to others are still the subject of heated scientific debate. I also realize that in order to learn what is presently known about any of them I need to go back to the same source: Sirpa Hakkinen.

For several months after my initial meeting with Hakkinen at Goddard, I'd been forced to put our encounter on hold while I tried to learn what I could about the GSA. Finally, however, I felt ready to make another journey to meet with her. In the intervening months I had perused the Dickson study and had become familiar with both the idea and the history of the GSA. Now I wanted to learn more about Hakkinen's mathematical models of the ice-ocean boundary and what they might reveal about the causes, the behavior, and the long-term effects of a freshening event such as the GSA.

Hakkinen began this meeting by talking about the whole

idea of mathematical modeling as a strategy for studying the Earth's climate system. One of the most useful things about such a model, she explained, was that it could be used as a kind of experimental testing ground for ideas about the real world—ideas such as Dickson's advective hypothesis about the origin and behavior of the GSA. Dickson had made a rough test of his idea by comparing a few scattered measurements of the event's size and speed with a set of hypothetical values in a best-guess scenario. A mathematical model such as Hakkinen's could provide additional tests that were far more specific. If the inputs at the beginning of the model run were approximately correct, the model could be expected to imitate actual events and produce outcomes that were similar, if not identical, to those observed in the real climate system.

I asked Hakkinen if her model verified Dickson's advective hypothesis for the GSA.

"Indeed it does."

"And after that—does it reveal anything more? For instance, does it give us any idea about what might have caused the GSA in the first place?"

"That's a tricky question," said Hakkinen. "If you're asking about primary causes, such as whether human activity might have had something to do with it because of the way we've been changing the chemistry of the atmosphere, the answer is that it's still too soon to tell. Nobody in the scientific community is ready yet to speculate about the primary causes of an event like this.

"If you're asking about what might be called 'intermediate' causes, there are actually several proposals currently on the table. The one I think makes the most sense was first suggested in the late 1980s by a pair of researchers named Aagaard and Carmack. According to these two, the source of the anomaly was almost entirely Arctic, occurring in the form of a huge migration of fresh surface water and multi-year pack ice that flowed southward from the Arctic Ocean, through the Fram Strait [between Greenland and Iceland], and into the Greenland Sea.

"According to Aagaard and Carmack, the reason for this event was a change in what scientists term the 'wind curl' over the eastern Arctic—a change caused by a somewhat mysterious increase in mean atmospheric pressure over the area between northern Greenland and far northern Europe."

"So for some unknown reason the weather pattern changes—and a mass of freshwater and pack ice floods down through the Fram Strait and into the Greenland Sea," I said. "Then what happens? How does all this extra ice and fresh-water affect the operation of the Great Conveyor Belt, for instance?"

Here, Hakkinen suggested, we needed to look more closely at her model. The two runs that she'd done over the past several years had both correctly predicted two specific outcomes at the ice-ocean boundary. First, the model had shown that a massive freshening of the ocean surface in the form of ice and seawater from the Arctic would also condition the surface of "downstream" areas, resulting in increased sea ice production across the region. And second, the model had indicated that a large Arctic freshening event would also "cap" the surface of the Greenland Sea, causing stratification of the water column and a slowing or stopping of deep convection.

"In fact," Hakkinen remarked, "actual events during and after the GSA have confirmed both these model behaviors. Sea ice extents in the Greenland Sea have indeed increased after every major ice export event. And later oceanographic measurements have indicated that deep convection and the production of Greenland Sea Deep Water have either been greatly reduced or have stopped altogether since the late 1970s."

"Even so, the GSA doesn't seem to have shut down the Great Conveyor Belt," I said. "Europe is still a temperate climate. There aren't any glaciers growing in Canada. The Gulf Stream is still operating on schedule . . ."

"There seem to be a number of responses that the Conveyor system can make to changes in the rate of deep water formation—both in the Greenland Sea and elsewhere," said Hakkinen. "Along with Broecker, I suspect the system may have a number of stable modes of operation. Some of these modes seem to be controlled by a kind of on/off 'switch,' while others seem to respond to a sort of reostatic 'dial' that will slow down or speed up the Conveyor while allowing it to continue its overall operation."

"You mentioned that the production of deep water in the Greenland Sea has been slowed or stopped ever since the late 1970s. I'm not sure I understand why this slowdown has lasted so long. According to Dickson, the GSA moved out of the Greenland Sea by the late 1960s and didn't return—at less than a fifth of its original size—until 1982. How does your model account for the absence of deep convection in the Greenland Sea for all the intervening years?"

"The GSA was probably the jolt that capped the surface layer and got the whole slowdown started," Hakkinen explained. "But there have also been several other large pulses of sea ice and freshwater that have advected through the Fram Strait besides the one in 1968. The most recent of these events that I've included in the model took place during the two winters of 1981 and 1982, at about two-thirds the magnitude of the GSA. That one may have affected deep convection and the production of deep water in the Greenland Sea for most of the 1980s."

I became silent for a moment, pondering the magnitude and intricacy of the geophysical events that Hakkinen was describing. "The whole business is mind-boggling," I said at last. "There are so many variables, so many inputs, so many possible responses. The complexities of this system must stress your mathematics . . ."

Hakkinen stared at me for several seconds, then leaned back in her chair and laughed. "The complexities of this system stress your *imagination!*" she retorted.


The evening grows late and the light outside *Brendan*'s cabin begins to fade—an indication of how quickly the summer is passing. The noise of wind in the rigging has dropped several decibels since the afternoon and the rain has tapered off to a fine mist—both signs that the gale center has moved up the coast and that a clearing is on its way.

I have one more question to sort out this evening, and that is what to make of the heavy influxes of winter sea ice in the waters around Manitsoq that the fish plant manager, Peter Beck, described to me this morning. With what I've learned from Hakkinen, there seem to be a number of possible explanations for such occurrences. One explanation would be to assume that we are now in the midst of another event similar to the GSA and that the winter ice along this coast is a result—ice that has been carried here directly out of the Greenland Sea.

Another explanation hinges on the proposition that, at least partly as a result of the GSA, the West Greenland Current has slowed down, causing water surface temperatures to fall all along this coast and providing the potential, at least, for additional sea ice formation here in western Greenland.

A third explanation has to do with deep convection. Why should the Greenland Sea be the only place where deep water formation has slowed or stopped? If the circulatory system of the northern Atlantic has indeed become more sluggish since the time of the GSA, then maybe deep convection has slowed and a layer of fresh surface water has also formed across portions of the LDB area—just as it did in the Greenland Sea. One possible result would be larger sea ice extents around this section of the northern ocean, too. This might help explain the local sea ice anomalies in Greenland during the last couple of winters. It might even help explain the heavy summer ice along the Labrador coast that *Brendan* and her crew ran into in 1991.

As I stretch my mind farther and farther around the intricacies of the ice riddle, I find myself shaking my head in a kind of befuddled wonderment. There are so many dimensions to this riddle—the existence of advective events such as the GSA, the processes of deep convection, the concept of the Great Conveyor Belt, the notion of "switches" and "dials" that govern its operation. Hakkinen was right—the complexities of the system do indeed stress your imagination!

*August 8, Monday. Talerulik Island, Greenland.*

The heavy seas from yesterday's gale recede slowly as *Brendan* resumes her journey south. For six hours we wind our way behind barrier islands and meander down long, protected sounds. In the late afternoon we round a steep headland and approach the last protected anchorage along this outer passage for the next hundred kilometers—a blasted lump of mottled rock called Talerulik Island.

As soon as the anchor is set, Mike and Blue launch the rubber boat for a trip ashore. Just before they cast off, Pete arrives on deck carrying a satchel of collection jars, asking if he may join their party. "I'm going to perform a Darwinian census," Pete announces to no one in particular. "I'm going to collect one specimen of every living thing that grows on this island."

Earlier in the summer I might have laughed at this remark. But today is not the first time Pete has set off on such a venture. His cabin, in fact, has been filled to overflowing on any number of occasions with container after container of tundra flowers, mushrooms, pitcher plants, crowberries, lichens, mosses, Arctic grasses. There was a day several weeks ago when I thought there'd been a bloom of some nasty smelling algae in the bilge—or an explosion of rotting potatoes or onions or cabbages in one of the food lockers. I was ready to set the crew on a boatwide search for the offending substance until Pete finally confessed that it was his cabin— or rather the collection of specimens he was storing there— that was the culprit.

Pete looked ready to cry when I told him that his rotting floral samples would have to be tossed overboard. "Where would the world be today," he asked, "if the *Beagle*'s skipper had told young Darwin that his samples were too smelly to take home?"

In response to this somewhat daunting rebuttal, I finally modified my request, indicating that he could store his samples in sealed plastic food containers in the forward sail locker (all except the mushrooms which, as far and away the worst offenders, would still have to be pitched overboard). The forward sail locker has smelled like a boatyard dumpster ever since.

In spite of the barrenness of our present setting, I suspect Pete may have difficulty living up to his promise to collect one of every species that grows here. Even in such a blasted landscape, ravaged by storms and scoured by ice, there exists a surprisingly large and tenacious biota. On the west-facing (windward) side of the rock almost nothing grows. But on the east-facing side there is, as Pete has observed in his journal, "an amazing array of low, scrubby, mossy, licheny, cactusy stuff." "Incredible!" he remarks in another entry, written soon after his first foray ashore in Greenland. "This place is totally alive!"

Amanda and I remain on board while the others walk on the island and Pete collects his samples. While we wait, we begin preparations for what will soon become the evening's main activity: a roundtable session in which Pete will be master of ceremonies, leading his shipmates in an investigation of the samples he has found. Together we will peer at mushroom spores and watch copopods swim in random circles under the microscope. We will look up their names in field guides and talk about their roles in the Arctic ecosystem. We will reflect on the interdependency of the life web here, on the small number of species, on the surprise of there being communities of biomass existing in this place at all.

As Pete once observed, we'll be doing what many of our fellow human beings also do each evening—watching a home-produced version of "Greenland TV." The difference is that the characters in the program we watch will be distinctly nonhuman, the plot will be a little thin, and the message will have more to do with what the natural world is trying to tell us than with what we might prefer to hear.

*August 11, Thursday. Nuuk, Greenland.*

The sounds of automobiles and motor scooters and delivery vans echo off the hills above Nuuk harbor. A bell rings somewhere in the town, and a factory whistle answers. At the end of the commercial quay a large Royal Greenland stern-trawler offloads palates of frozen shrimp, while motorized lift-trucks move them onto the deck of a container ship that waits on the far side of the quay.

It's odd—but ever since *Brendan*'s arrival in this harbor the day before yesterday, I've felt a jarring sense of dislocation, as if we've been transported to an urban setting somewhere far to the south. The town is noisy and crowded. The air is filled with the smells of gasoline and heating fuel. Garbage floats in the mooring basin. Suddenly the pristine landscapes to the north seem far away, and I feel disoriented and out of place.

This morning, however, there is no time to dwell on these feelings. As my crew and I begin preparing for the passage across the Davis Strait, I need to turn my attention once again to the ice reports and weather forecasts for the area we are about to traverse. I know the sea ice has been receding along the Labrador coast during the early weeks of August. Now a look at the most recent ice reports shows that the ice along the coast of southern Baffin Island is also clearing for the first time in several years. This means that the rhumb line course—direct from Nuuk to northern Labrador—should carry us some seventy-five kilometers south of all reported ice for the entire nine-hundred-kilometer passage.

The news from the weather forecast maps is also good. The storm track that crosses Labrador and swings north into the Davis Strait has been highly active over the past several weeks. One system passed here several days ago while *Brendan* waited in Manitsoq. Another system veered south yesterday and is crossing southern Greenland today. A third system—the strongest of the three—is currently forecast to arrive in northern Labrador in about five days, or sometime next Tuesday. This means that if *Brendan* can be ready to sail by this evening, she will be looking at a window between weather systems of approximately four days—plenty of time, I tell myself, for the passage we are contemplating.


At twenty minutes before twelve, in the darkness of actual night, Mike and Amanda slip the mooring lines, I drop the engine into forward, and *Brendan* moves slowly past a row of fishing boats into Nuuk harbor. The night is inky black, and the visibility drops to nil as the sailboat pulls away from the lights of the town. Beyond the bar at the harbor mouth Pete raises the mainsail and Blue moves to the bow to watch for growlers and bergy bits that we have noticed during the last few days floating in the fiord.

The wind rises. *Brendan* accelerates into the night. The navigational light that is supposed to show from the northern corner of Agtorssuit Island is extinguished, so I put *Brendan* on a compass course and feel our way past the last few dangers at the mouth of the fiord. Blue calls from the bow each time he sees a slab of floating ice. Soon I notice that I, too, can just make out the shape of an occasional white object low in the water, and I realize that the dawn is starting to grow behind us.

Ahead there is no more land—only the gray backs of seas that are already starting to mount and break. I know that if I look around, I will see the mountains of Greenland rising in a jagged line against the morning sky. But I do not look around. I don't want to think about leaving this coast but only about the stark, mountainous coast on the other side of the Davis Strait that has filled my imagination for so long.

# VII.

# Southwest

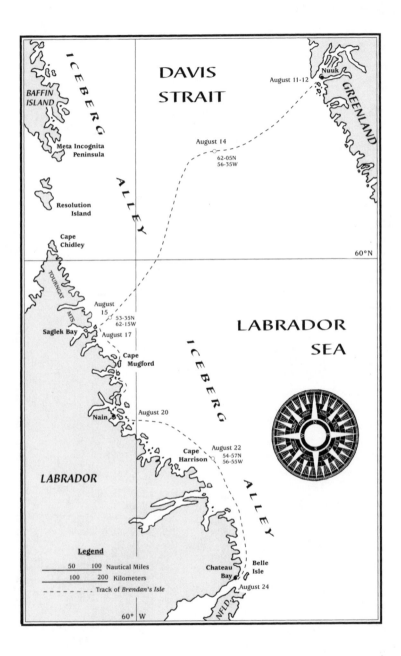

BAFFIN
ISLAND

ICEBERG

DAVIS
STRAIT

August 11-12    Nuuk

GREENLAND

Meta Incognita
Peninsula

August 14
⊙
62-05N
56-35W

ALLEY

Resolution
Island

Cape
Chidley

60°N

TOURNGAT
MTS

August
15
53-35N
62-15W
Saglek Bay
August 17

LABRADOR

SEA

Cape
Mugford

ICEBERG

Nain
August 20

August 22
Cape        54-57N
Harrison    56-55W

LABRADOR

ALLEY

Legend

50     100  Nautical Miles
100    200  Kilometers
- - - - - - . Track of *Brendan's Isle*

Belle
Isle
Chateau
Bay
August 24

NFLD.

60° W

For twenty-four hours after leaving Gotthabsfiord, *Brendan's Isle* races before strong easterly winds, skidding down the faces of following seas, vibrating and shuddering, throwing a curtain of spray from her bows, carving a line of white foam in her wake. The knot meter ticks ten, eleven, twelve knots as she surfs downhill, surrounded by streaks of spindrift and flanked by long, breaking crests.

On Friday evening the wind eases back and veers into the south. It drops lighter and lighter as the dawn breaks on Saturday, until at last the surface of the sea grows glassy-smooth and *Brendan* moves like a ghost ship in almost total silence. Fulmars and kittiwakes soar in semicircles in her wake. An occasional iceberg turns a slow pirouette on the horizon, reflecting light and shadow off its faceted surfaces as it moves northward in the current.

Soon after breakfast on Sunday a series of weather forecasts breaks the serenity of the passage and forces us to confront a difficult decision: a choice between unwanted alternatives. I've learned from several weather-fax reports that the large area of low pressure that was due to arrive in Labrador in two more days has started to move faster and that the system is deepening, with wind forces now being described as north gales forty-five to storm fifty-five knots.

That's too much wind. *Brendan* does not have the ability to make safe landfall under such conditions on a mountainous and largely uncharted coast. She must either run through iceberg alley in the dark and at speed tonight in order to raise the coast and be safely anchored by early tomorrow morning—or she must stop where she is on the eastern side of the ice, heave-to, and ride out the storm at sea.

The most prudent move would almost certainly be to remain at sea. A rockbound and poorly charted coast presents far more danger to mariners than almost any storm—and as long as *Brendan* has sea room, she can run south before the wind with little risk to herself or her crew. The one drawback to this alternative is that in a large system such as this one, she might have to run halfway down to the Strait of Belle Isle before the storm blew itself out, and thus, just as in 1991, she would once again miss visiting the coast of northern Labrador.

The second alternative—running at speed through the ice tonight—is equally undesirable. I know that there will be five hours of darkness in these latitudes, and the moon is still new. *Brendan*'s fiberglass hull is, of course, not ice-strengthened, nor does she have a crash bulkhead in case of holing in her bows. The area of bergy water that we must cross is a hundred fifty kilometers wide, and we'll need to maintain a speed of at least six knots through the nighttime hours to ensure a timely arrival on the coast by tomorrow morning.

Maybe if I didn't know that *Brendan* carries a good radar and several young crew members with excellent night vision, I would veto my shipmates' unanimous desire to run through the ice. But I do know these things and therefore agree.

We'll keep three on watch at all times tonight, I decide, and we'll rotate a radar operator and a person on the bows every half hour, adding an extra lookout forward if we get into heavy ice conditions. We'll run at speed directly for Saglek Bay—a fiord in the Tourngat Mountains with a decent anchorage near its mouth that is protected from the north. And we'll cross our fingers, rub every luckystone, and kiss every four-leaf clover we can find—hoping we make it safely through.

*August 15, Monday. 58/35N, 62/15W.*

*LOG: Wind: northeast, 20 knots. Barometer: 29.17 falling rapidly. Weather: heavy overcast. Position: iceberg alley, Labrador Sea, 15 miles due E of Saglek Bay, Labrador.*

At three-thirty the night is still dark. The wind has started to rise from the northeast. The only sounds in *Brendan*'s cabins are the rush of water along the hull, the whine of wind in the rigging, the hum of the radar as it pulses around and around, casting an eerie glow on the face of the person who watches.

I've just been relieved after my half hour's duty on the bow, so at the moment I'm the only person on board who is not on watch. I lie on my back on the settee in the main saloon, fully clothed, listening to the sounds of the wind and feeling with my body the motion of the boat as she rolls and plunges through crossing seas, drawing ever closer to the Labrador coast.

It's funny, I think, how easy it's become for me over the years to distinguish among the various components of the crossing wave train as *Brendan* moves through them. There is the short, chattery motion of the northeast wind-wave slapping against the hull and bow, mounting as the wind mounts. There is the slow rise and fall of the southerly groundswell as it marches up from a weather system somewhere far down in the Atlantic. There is the pitch and yaw of Saturday's easterly—a last, dying memory of coastal winds far astern. And now there is another motion recently added—an increasing surge from the north, angry and insistent, that serves as a warning of more wind to come.

A dozen years ago, before I knew *Brendan* as well as I know her now, such motion at sea had often seemed random and chaotic. Rolling and jerking and slamming into waves, the boat had seemed a thing possessed, and on the roughest nights I'd even felt a twinge of nausea.

But now the motion is comprehensible. The sounds that the hull makes as it moves through the water are like the whisperings of an old friend. The various combinations of surges and plunges and long, slow rolls are like the rocking of a long-remembered cradle. The wave train, rather than seeming an alien thing, has come to feel like a familiar messenger, carrying news from half a dozen sources and delivering it together in a complex chorus of sound and motion.

How is it, I wonder, that the human brain is able to sort through such a mishmash of intersecting patterns and somehow find the order in them? Ocean waves, after all, are highly irregular. The series of crests and troughs that they produce as two or more crossing trains intersect create a pattern of the type that mathematicians term "nonlinear"—that is, a pattern that refuses to repeat in an orderly manner but only continues to permutate in endless new combinations.

It seems that the human brain must have the ability to collect and store such complex information in a plastic way so that we may shuffle it, sort it, play with it, recombine it, subject it to the most fantastic leaps of imagination. The unedited patterns—like those of the crossing waves—may at first seem chaotic, yet after a time the brain tells us that they are merely complex, that indeed they do contain logical groupings, and that these groupings are finally comprehensible.

The deep ocean is not the only setting in which one may develop this capacity for dealing with complexity. We are surrounded everywhere in our experience by complex systems. The rise and fall of the wind as it blows across an open field, the movement of clouds in an approaching thunderstorm, the dance of flames in a fireplace, the flow of traffic on a crowded freeway, the pattern of branches in a forest in winter, the behavior of gases at the surface of a boiling pot of water—these and thousands of other phenomena in the world of everyday experience exhibit the qualities of complexity, and most of us learn to think intuitively about them and eventually to make reasonable sense of them.

Our traditional mathematics and classical sciences have always had difficulty with phenomena such as these, preferring instead to focus on patterns that seek closure, problems that have single answers, equations that can be "solved." But the new cybernetic sciences, armed with powerful computers, have embraced complexity. Indeed, every one of the scientists whom I have contacted over the past several years with regard to the ice riddle has been intimately involved in the study of complex systems. Parkinson's satellite-sensed data sets of sea ice growth and decay, Marko's correlations of side channel bridging and Baffin Bay sea ice cover, Broecker's convective processes for the production of North Atlantic Deep Water, Alley's compilations of ancient snowfall amounts as they relate to changing patterns of climate, Hakkinen's model of deep ocean mixing at the ice-ocean boundary—all exhibit the same involvement with complexity.

Like sailors learning how to read a crossing wave train,

each of these scientists has needed to become familiar with the seeming randomness of the system he or she is trying to understand. Like sailors, they live with the systems day after day. They manipulate them, model them, feed them into computers and "run" them. Sometimes when nothing else seems to work, they simply store them in some out-of-the-way compartment of the brain and let them sit there unattended for a while. Eventually, with time and patience, they begin to sense emerging patterns and to grasp the intricate logic of how the various components fit together.

There is a sense, I think, in which this capacity for dealing with complex systems may simply be a matter of temperament. There are those who naturally seek out situations that are orderly. And there are others who seem drawn to process and pattern and to the inherent messiness of nonlinearity—people who say "ah, what a lovely, complicated business this is" and then just roll up their sleeves and dig in.

There is another sense, however, in which our ability to deal with complex systems may be growing more commonplace—a result, some have suggested, of an emerging "worldview of complexity." A century ago almost everyone in Western society thought in terms of the old equilibrium viewpoint—the idea that there is a fundamental duality between humans and nature and a point of natural equilibrium between them.

But increasingly today there is another way of understanding ourselves and our relationship to nature—a kind of interactive worldview in which there is no longer a division between ourselves and our natural surroundings. Instead, both are seen as components of the same complex system. One does not reside outside or beyond the other, but rather the two are integrally connected in a single, continuous, and highly interactive relationship.

What this means, among other things, is that as human beings we can now begin to think of ourselves as part of what may cause natural systems to change their mode of behavior. When we take a collective action in response to a particular situation in nature, we set in motion a sequence of events that may eventually come back and *create a different situation for us to adjust to.* There is no point of stasis. There is no "natural equilibrium." We are part of an ongoing process, and as such, we have the ability to provoke the system into any number of new and unpredictable responses.

It is this interactive worldview, I think, that informs many of the new sciences of complexity (such as the study of Earth climate) and that gives rise to statements like those of Wally Broecker about the "coming surprises" that we seem destined to experience as we continue to perturb the climate system. As Broecker admits, we really don't have any idea yet what these surprises may be—for the system that will generate them is complex, and the range of possible responses is still beyond our ability to predict.

But just as a crossing wave train will eventually result in a square-shaped sea, a tumbling crest, a rogue wave, so the climate system will eventually respond in dramatic and unforeseen ways to the perturbations that we collectively visit upon it. As we continue to multiply in numbers, as our waste stream increases, and as our technologies grow ever more powerful, we begin to behave like another train of waves marching across the surface of the sea—a strong and growing force interacting with all the other forces. We are not separate. We cannot escape our involvement with the patterns that are generated. As we add our own inputs to the process, the patterns will change, and we will be required to respond to new patterns that we ourselves have helped to create.


The ship's clock chimes eight bells—four o'clock—and I realize that I must return to the deck. I climb the companionway ladder to find Mike at the helm and Pete sitting next to him in the cockpit. At first I wonder why Pete isn't watching from the bow—then I realize that in the half hour I've been below, the dawn has begun to break and the light has already started to change.

A crimson glow grows under a layer of overcast behind us. The vague form of an iceberg looms on the horizon several miles to the west, and another appears dead astern framed against the rising light. I relieve Mike at the helm and we sail in silence for the next twenty minutes as the light grows—an angry color that washes the sky overhead with streaks of blood red and that casts an eerie glow on the surface of the sea.

There is no doubt now in any of our minds about the approaching weather. The boat heels to a gust of wind, and I strain ahead for a glimpse of the coast that we must find before the gale arrives.

*August 15, Monday evening. Eastern Harbor, Saglek Bay, Labrador.*

The wind screams in the rigging. The boat shudders and groans. The noise makes it impossible to talk in the cabins without raising your voice to a shout. Outside, the rain blows horizontally, driving against ports and hatches and forcing itself into every crack. *Brendan* lies to two large fisherman anchors in the northern end of an uncharted cove, protected from the seas by a low spit of land but wide open to the force of the wind. Mountainous terrain on either side of the cove funnels the gusts in a venturi effect. An open stretch of water a few hundred meters to the west is whipped into a lather of steam and spindrift by winds that have been accelerating all afternoon to hurricane force and above.

I sit at the nav-station as darkness falls, listening to the sounds of the storm and thinking about our situation here. The wind shows no signs of relenting—in fact, it feels as if it may be blowing harder now than it had been earlier this afternoon. The barometer is still falling. Last time I looked, it had dropped to 28.7 inches of mercury (972 millibars)—lower than many hurricanes ever get.

There are any number of reasons that I should probably be feeling anxious. The keel could ground on an uncharted rock. The anchors could fill with kelp and begin to drag. An anchor line could chafe through. A crew member could get sick or hurt. The wind could shift and blow us onto a ledge at the side of the cove—or it could veer a hundred eighty degrees and pin us against a lee shore.

For some reason, however, I don't feel particularly anxious about any of these things. I'm aware of our vulnerability here and of the possibility of all sorts of sudden emergencies. I realize that we need to remain vigilant and that we must be ready, if any of them should materialize, to act quickly. But meanwhile I feel a kind of tranquility in the face of this storm—not so much anxious as awed, not so much frightened as overwhelmed by the magnitude of the events around us. I feel much the same as I remember feeling in other weather emergencies at sea: a six-day gale south of Iceland, a late-summer hurricane in New England, a black squall in the Gulf Stream, a wall of waterspouts in the Sea of Alboran, a storm in the Cabot Strait.

Such events *can* be frightening as they actually take place, for they speak to us of our mortality. They remind us of how helpless we sometimes are in the face of an indifferent nature. But they are also acceptable somehow—because they are a part of our age-old situation in nature and because they are beyond our control.

What is to me much more frightening—what is, in fact, truly terrifying—is the thought of an occurrence like this that is not merely an event of nature. A storm, let's say, that no longer behaves according to the patterns we have come to expect over the centuries—a storm that is a "surprise," a "nonlinear response" caused by human-induced changes in the patterns of climate.

Speculation about what the future might look like as the

Earth begins to respond to $CO_2$ doubling or other green-house perturbations includes all sorts of unexpected climate events: anomalous increases in the number and magnitude of hurricanes, larger and deeper extratropical storm systems, stronger winds, denser cloud cover, heavier precipitation, longer droughts, abnormally large changes in sea ice cover, sudden growing or melting of polar ice caps, catastrophic changes in global sea level.

The thought that nature may not be nature any longer—the thought that the human species may now be sitting at the controls and that our collective actions may be causing events to occur in nature—this is a bone-chilling thought. There are those who talk about "managing the planet"—as if humankind were actually capable of taking over and somehow operating the climate system to its own advantage. The arrogance of such a notion is superseded only by its foolishness—for to imagine that any of us is prepared to predict how the system will react to our tampering is to misunderstand the nature of complex systems and the unpredictability of nonlinear responses.

No, I think to myself, give me raw nature any day: a hurricane, an earthquake, a North Atlantic storm. I'll take my chances with one of these a hundred times before I'll assent to intervention—willful or otherwise—by a species as careless and shortsighted as our own.

*August 17, Wednesday morning. Eastern Harbor, Saglek Bay, Labrador.*

The storm rages all Monday night and most of the day Tuesday. Finally late Tuesday afternoon the wind begins to moderate, and by early evening the rain stops and there are a few ragged breaks in the sky to the north. Amanda climbs up on deck to fill her lungs with fresh air and stretch her legs. She looks longingly at the beach. Then she asks if she might retrieve the rubber boat from where it has been stowed belowdecks, inflate it, and launch it for a short trip ashore.

I feel like an ogre as I veto her request. The wind is still gusting into the thirties, I remind her, and there's no guarantee that it may not come around and begin blowing again from some other direction. *Brendan* isn't out of danger yet—she could still become trapped here on a lee shore. It's not time to have the rubber boat on deck where it might get in the way if we have to do some quick readjusting of the anchors.

Late Tuesday evening the wind moderates again. I ask for hands to help shorten the anchor lines, then I set a schedule for an all-night anchor watch. I'm concerned now that when the wind does finally drop, a leftover swell might turn the boat around and carry her onto the rocks. "We'll each take an hour at the radar during the dark hours," I say, "so we can keep an eye on our position. I'll take the first hour, from eleven to midnight. Amanda will take the last."

After thirty hours of incessant noise, the silence when the wind stops is oddly unsettling. Once my hour on watch is done, I climb into my bunk, pull my sleeping bag up around my ears, and stare at the overhead. I don't fall asleep for a long time—I have the feeling that I don't fall asleep at all. But eventually I must, for suddenly I'm aware of someone standing next to me, shaking me by the shoulders.

"Skipper, wake up . . . wake up." Amanda's voice. "Skipper, there's something out there—something you need to see."

I sit up too fast and knock my head on the lamp above the bunk. My mind begins to race, filled with images of *Brendan* grounded on a ledge or pinned against a rocky shore. "Huh? What is it? What is . . ."

"Don't worry, there's nothing the matter with the boat. I just think you need to come up on deck and see what I've been looking at."

I pull on my Mustang coverall and a pair of boots and follow Amanda through the galley area and up the companionway ladder. For the next several seconds I am utterly blinded by the darkness that surrounds the boat.

"What's going on?" I ask again. "What is . . ."

Amanda leans over and covers my mouth with her hand. "Shhhhh . . ." she warns in a whisper. "Don't make a sound. You may frighten them away."

She points into the darkness to the left. I stare, blink, rub my eyes. Then slowly, as my pupils adjust to the faint glow of the dawn just breaking, I'm finally able to distinguish a cluster of white forms—moving—about twenty meters from *Brendan*'s cockpit.

"Polar bears," Amanda whispers. "A mother and two cubs. She's been stalking back and forth on that flat section of rock for about five minutes now—trying to figure out who we are."

We watch in silence as the dawn rises. The adult bear continues to stalk, lifting on her haunches from time to time to sniff the air and stare out at the sailboat. She knows we are here—she has our scent, and now she can see us. When *Brendan* swings sideways in the swell, our stern is carried toward the shore until we are only about a boat length from where she stands. She looks straight at us, utterly unafraid, while her cubs wrestle and tumble at her feet.

Eventually she seems to tire of our company, and she turns and begins to forage among the boulders. The cubs follow, circling close, as if they are connected to her on invisible tethers. She appears to move slowly, yet she covers the ground with surprising speed. One minute she is standing on the ledge a few feet away from *Brendan*'s cockpit; the next minute she has traversed several small hills and is moving across the beach at the head of the cove.

I glance at Amanda and smile to myself, suddenly happy that there was too much wind last night for her trip ashore in the rubber boat. The encounter we've just had this morning feels like a gesture of truce from this wild land after our ordeal of the past two days. But I know it could easily have been otherwise—and I'm thankful that *Brendan*'s crew are unharmed and ready to continue on.

*August 20, Saturday. Nain, Labrador.*

Ever since the gale stopped blowing four days ago, there has been no wind on this coast. The passage from Saglek Bay to the Port Manvers area was all done under power. And to-day—once again—the engine remains on for the entire sixty-kilometer run along a winding inner channel toward the settlement of Nain.

By the time we round the last headland and see the buildings of the town ahead, I know I'm in trouble with Blue. But I don't know just how much trouble until we pull up at the government pier and begin the job of setting up the mooring lines.

Blue climbs from *Brendan*'s bow to one of the creosote pilings of the pier, then onto the wooden deck above. He takes a line from Pete and begins carrying it to a large bollard forward of the bow.

"Bring that line aft!" I holler at him. "And don't tie a bowline—just throw on a couple of hitches and come back for this one."

At the sound of my voice Blue stops and turns. He doesn't say a word but only stares, with the veins bulging out on his neck and a flush of emotion growing on his face. His hands begin to tremble, and he lets the line drop to the ground.

I ask Pete to climb to the pier to give his shipmate a hand. Then I look back at Blue, directly into his face, and I point at the piling where I want the line attached. Our eyes meet, and instantly I realize that it's not just my impatience over the mooring procedure that's causing the problem this morning. It is also this other thing—this silent battle that we've been having for over two months. It is the war of the diesel engine, and I know that somehow, for the good of this whole enterprise, we must try to do something to resolve it.

Once *Brendan* is properly moored, I climb to the pier and invite Blue to take a walk with me out to the end of town.

"We need to talk," I say.

"I guess we do," says Blue.

We walk in silence for several minutes, past the Moravian church, past the boardinghouse and the community center, along a dirt road that circles the harbor and ascends into the hills beyond the village. I begin by telling Blue how much I like him—how good he is at everything he does—how valuable he's been to this crew and to this undertaking. I know he needs to hear these things—and I know, too, that I haven't said them to him often enough this summer.

Blue smiles and seems to relax, and I change the subject. "Diesel engines," I say. "We need to talk about diesel engines—and this war we've been having over them."

"What's to talk about?"

"Hmmm . . . what's to talk about? Well, for one thing, you've made me think. You've really made me *wonder*, since I basically agree with the message you've been trying to send with all your heckling, why haven't I given in? Am I just being stubborn? I mean *why,* every time there's been no wind, have I kept on turning on that engine?"

"Good question," says Blue.

"The answer's so obvious that for a while I couldn't see it. But the fact is, Blue, that engine has been built into the structure of this voyage from the very beginning. It's been a *precondition* from the moment we left the Chesapeake Bay. It's the reason we've been able to keep our promises and do what we've planned to do all summer."

"John Davis sailed the same route—three times—and he didn't have an engine."

"That's true. But he didn't have the same kinds of constraints that we've had, either. He didn't have anybody to meet in Woods Hole or Sisimiut or Illulisat. He didn't have a three-month window for getting Pete and Mike back to their jobs and Amanda back to graduate school. Those are things that have been givens for us—they're a priori conditions, built into the plan."

Blue shrugs. "Then maybe we shouldn't have a plan. I mean, what's more important—making a rendezvous with a schooner in Illulisat or poisoning our planet? *Brendan* burns a gallon of fuel every hour—that's twenty-two pounds of carbon dioxide added to the atmosphere for every six miles she travels—almost four pounds a mile. If we could just . . ."

"Wait—*stop!* I *know* those things. And I *agree* that they're a serious problem. But I'm afraid you may be asking for a lot more than you realize. Think about the preconditions, Blue—the a priori's! They're not only built into this voyage. They're also built into all our lives. They're built into the whole invisible fabric of our society.

"You're not a Thoreauvian man, Blue. None of us is. We all participate in the problem. We all rendezvous with schooners in Illulisat, one way or another. We all precondition our lives around the convenience and usefulness of fossil fuels and internal combustion."

Blue purses his lips and scowls. "But you're supposed to be doing something different. You're sailing to Greenland to try to raise consciousness about the buildup of greenhouse gases and global climate change. *You,* of all people, ought to be . . ."

*"I* of all people? I'm no different than you—or Mike or Amanda or Pete or anybody else. We're all caught in the same dilemma. We're all participants in the same system with the same inertias, the same set of assumptions, the same uninspected values.

The *real* challenge, my idealistic friend, is not just to turn off the engine in this sailboat once in a while and float around becalmed. That might be nice—but it will never be anything more than a gesture—something to make us feel good about ourselves. No—the real challenge involves something more basic—it involves *investigating the a priori's!* Taking a hard look at the *preconditions* that determine the choices we make and the ways we act . . . as individuals . . . as a society.

"The argument you and I have been having is pointless, really—because with the plan for the summer already in place, the option of eliminating the engine has never been a realistic alternative. If we'd wanted to have an argument, we should have been doing it six months *before* this voyage, and we should have been arguing about *the plan itself!"*

"I wasn't around six months before the voyage—I wasn't consulted when the plan was being made."

"No, you weren't—and I'm sorry about that, because I know you'd have added a valuable voice to that process. But I think you might also have learned something."

"What's that?"

"Well, for one thing, you might have learned that the

choices we face are much more complex than you seem to imagine—and that the solutions rarely involve a simple either/or. Yes, we need to learn to turn the engine off—that's true whether we're sailing to Greenland or just living our day-to-day lives as citizens of this planet. But we also need to learn how to operate the engine more efficiently, develop conservation techniques, devise safe technological alternatives. We need to learn to differentiate between the times when the engine may be necessary—even essential—and the times when it's merely convenient. We need to become aware of the risks of overuse, and we need to learn to weigh alternatives, measuring short-term gains against long-term costs.

"It's a complicated process, Blue—not just a matter of enforcing a simple yes or no. Confront people with a categorical demand, and you'll probably find yourself an embittered old man some day, raging against the stupidity of your own species. Engage in the difficult and complicated task of reassessing our collective needs and wants and reprioritizing our actions, and you may soon be helping to design solutions."

Blue nods, almost smiles. "You're not telling me anything I don't already know."

"Maybe I'm not, Blue. Maybe I'm just trying to tell you that I know them, too—and that I'd like to think of you as an ally instead of an adversary. There are too many problems out there looking for solutions for us to be fighting this battle—we're too close to being on the same side. And there's too little time left for quibbling.

"Back when I'd just graduated from college the human population of this planet was three billion people. Before the end of this decade it's going to be six billion, and by the time you're my age, if the demographic projections are even close to correct, it's going to double again to ten, eleven, twelve billion people. By that time you're not going to have to go around preaching to *anybody* about the impacts that our collective actions are having on this planet. The problems will be obvious for everyone to see. And the need for solutions won't be a matter for polite discussion—it will be a matter of survival.

"How about it?—can we call a truce?—talk when we disagree?—try to minimize the use of that engine whenever we can?"

We pause for a moment at the crest of a hill and look back down at the harbor and the rooftops of the town. Blue hesitates, then turns and meets my eyes with a steady gaze. "Truce," he says at last. "Here's my hand on it."

*August 22, Monday midnight. 54/57N, 56/55W.*

*LOG: Wind: northwest, 15 knots. Barometer: 30.08 steady.*
*Weather: clear. Position: Labrador Sea, 25 miles E of Cape*
*Harrison, Labrador.*

Amanda and I stand the dark watch tonight, midnight until
four, although with a waxing gibbous moon, there's been
enough light to cast hard shadows on the deck and illuminate the surface of the sea horizon to horizon. *Brendan* sails
before the wind with the jib poled out and the boat rolling
gently. Once again we're moving down iceberg alley, with
several silhouettes of ice shapes visible in the distance. But
with the cloudless sky and the bright moon and the decreasing number of icebergs as we travel south, there's not even
been a need for a bow watch tonight.

We are sailing along the southern third of the Labrador
coast now, closing the last of the Davis circle. Our original
plan had been to travel this final section of the coast using an
inshore route by day and anchoring by night. But the fresh
northwest breeze, the moonlight, and the clear weather have
been too good. With the wind astern, these are what sailors
call free miles, and we would be fools not to be taking advantage of them.

Ever since Blue and I had our talk back in Nain, the mood among all the crew has noticeably improved. Blue and I are communicating better on almost every level. The diesel war seems a thing of the past. Maybe all we needed was a clearing of the air. Or maybe the war wasn't such a bad thing after all—maybe we both managed to learn something in the process.

Mike is doing well, enjoying the rugged beauty of this coast, writing voluminous notes in his journal, collecting unusual rocks to give away to friends when he returns home. Pete continues his scientific collections, concentrating now on the flora of the Labrador tundra. Amanda just continues doing what Amanda does best—listening, maintaining a humorous patter, trying to keep things running smoothly among her shipmates. Tonight she seems quiet, however, and perhaps a little sad. She had visited this portion of the Labrador coast before, and she seems increasingly aware of our return to familiar waters and, with it, the inevitable winding down of this voyage.

The hours sound one after another on the ship's clock. The moon drops toward the hills in the west, and the night moves imperceptibly toward dawn. There are fewer icebergs visible around the horizon with each passing hour, and even as the moon approaches setting, there remains enough natural light on deck to read the compass without the aid of an electric lamp.

These are narcotic miles. There are no other human beings here, no ships, no aircraft, no lights along the coast, no sounds except those that *Brendan* herself makes as she moves through the water. Amanda and I exchange places at the helm and the nav-station every half hour, each of us immersed in the solitude of our own thoughts and the heady silences of a perfect night at sea.

*August 24, Wednesday. Chateau Bay, Labrador.*

On the third morning of the passage I discover a pre-dawn surprise: a navigational light flashing from a tower on the shore. Soon afterward the funnel of a ship appears in silhouette against the horizon ahead. Together, these signal the end of *Brendan*'s transit of the Labrador coast and her arrival at the eastern approaches to the Strait of Belle Isle.

The weather is clear and the wind is moderate—both good reasons for continuing through the strait this morning. But everyone on board is tired and in need of a stop, and a few kilometers ahead is a harbor that we are all familiar with—Chateau Bay. This is the harbor John Davis used as the rendezvous for the conclusion of his circumnavigation of this area in 1587. It is also the place where *Brendan* encountered the anomalous ice in 1991 and the place I began my deliberation of the ice riddle this past July. As such, it feels like an appropriate setting for us to close the Davis circle.

*Brendan* moves through the narrow passage at the mouth of the harbor and pulls up in the cove in front of the abandoned outport settlement. Once anchored, we launch the rubber boat and run in to the beach at Castle Island, there to stretch our legs after two and a half days at sea. I leave the others and make my way across the rubbly terrain to a standing rock at the island's southern end. I find a spot out of the wind where I can sit with my back against the rock, and I gaze out across the strait at the distant capes of northern Newfoundland, watching a solitary iceberg turn in the current.

Funny, I think. I came ashore, in part, to do a count of the remaining ice. And now it seems this single white triangle is the only object of its kind still visible. With August nearly over, even the late-season ice that often gets trapped in the mouth of the strait seems to have been swept out to sea, and I am looking at what may be—for me at least—the last ice of the summer.

Last ice . . . last ice . . . The phrase turns over in my mind like the descant to some sad, half-forgotten song. *Brendan*'s route, once she leaves Chateau Bay, will take her west—out of the Labrador Current and into the ice-free waters of the Saint Lawrence Gulf—then west again for two thousand kilometers to the Chesapeake Bay. After today we will leave the ice behind. Landscapes that have been littered with white all summer will feel strangely empty. The surface of the sea will appear bland and monotonous. Over the past weeks the ice has become a familiar presence—cold and silent and pristine—a reminder of the mysterious forces that move within the Earth. And I am going to miss it.

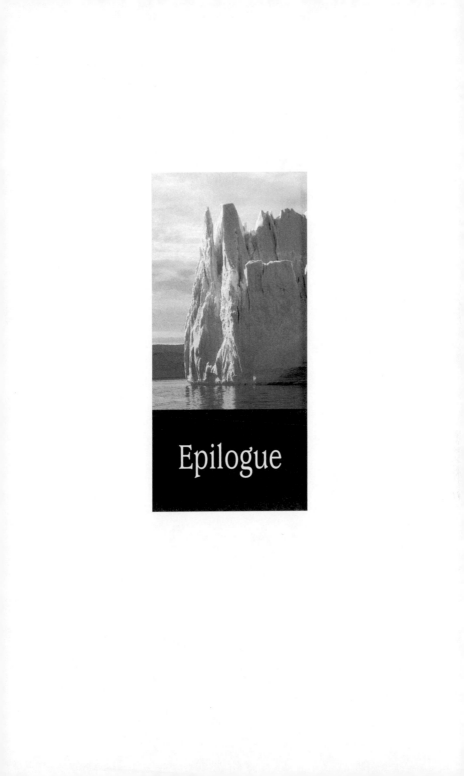

Epilogue

Three years have passed since *Brendan's Isle* and her crew and I have returned from Greenland. Except for a few tense moments during a gale in the Cabot Strait, our passage home from Labrador was uneventful. The most difficult part of that passage took place not at sea but on the land, after we had returned to the United States and were suddenly confronted with the culture shock of polluted estuaries, urban skylines, crowds of people, traffic jams. As often happens after such a voyage, we came home with our defenses down. For several days we felt like strangers, unnerved by the sheer numbers of human beings and by the consequences of their presence everywhere apparent in this landscape.

The culture shock didn't last long—it seldom does. For me, however, some of the aftereffects have remained. One reason has been my continuing encounter with members of the scientific community and with issues of changing global climate. The concern of almost everyone who studies the climate these days centers around the question of how much our collective human activity may be contributing to the greenhouse warming of the atmosphere. It is impossible to live with such a question for long without also being sensitized, the way any sailor might, to the impacts that our species is having on this planet.

Perhaps the most important event to have taken place in the world of climate study since *Brendan* returned home has been the publication of the *Second Assessment Report* of the Intergovernmental Panel on Climate Change (IPCC) in the winter of 1995–96. In this document, a group of twenty-five hundred researchers representing every relevant scientific discipline and nearly every nation in the world announced that they had found convincing evidence that the increases of carbon dioxide and other greenhouse gasses in the atmosphere correlate well with the global warming so far documented in this century. Furthermore, they concluded by consensus that there is now evidence of "a discernible human influence" on global climate.

The IPCC report has alarmed governments, galvanized environmental organizations, and focused the scientific community as never before on the importance of understanding the geophysical engines that drive the Earth's climate. There has been a flurry of research activity on dozens of fronts—but as yet, no dramatic breakthrough.

During the three years since I've been home, I have been able to talk again with almost every scientist who has appeared in this book—many of them several times. In most cases the stories they tell are continuations of the ones I'd learned from them before we sailed. In some cases, however, there have been important new developments.

The data from NASA's most recent generation of passive microwave imagers show a continuation of the 1978–87 decreasing trend in Arctic sea ice cover, although there has been disagreement about the magnitude of this trend. A team of Norwegian scientists has concluded that the rate of decrease has been accelerating, from 2.5 percent per decade during the nine years ending in 1987, to 4.3 percent per decade for the seven years ending in 1994. On the other hand, preliminary analysis by Claire Parkinson and her colleagues at NASA suggests a more uniform downward trend of around 2 percent per decade through the middle of 1995.

The bottom line in either case is that the Arctic sea ice cover continues to shrink. Meanwhile, the anomalously large ice extents in the Davis Strait/Baffin Bay area persist season after season. Canadian ice forecasters who have been monitoring this situation have also been keeping an eye on fast ice bridging in the Baffin Bay side channels. They report that the Smith Sound ice bridge has now either failed to form— or has formed late or sporadically—in five out of the past seven winters, fueling continued speculation that the southward flow of Arctic multiyear sea ice may be contributing to the LDB cold spot anomaly.

While occurrences such as these are worrisome to some observers, other events appear more hopeful. Recent evidence indicates that the convective slowdown that Sirpa Hakkinen and others had observed in the Greenland Sea for the past several decades has reversed its direction and that the production of Greenland Sea Deep Water has suddenly started again. A new generation of free-floating instrument platforms, called ALACEs, has been deployed in the Labrador Sea and elsewhere for measuring salinities and seawater temperatures in real time, and soon these measurements may help verify new theories about links between long-term atmospheric oscillations and massive freshening events such as the Great Salinity Anomaly.

The study of ancient climates and their relationship to present climate events also continues. In early 1995, Richard Alley and others published an analysis of the GISP II ice core data, comparing their findings with those of the European GRIP team from a year earlier. According to Alley, the two drill cores revealed the same timetable of events back to about one hundred thousand years, after which both records appeared disturbed by movements in the ice sheet. Thus the story of sudden climate change that the GRIP team thought it had discovered during the Eemian warm period will have to be put on hold while a third ice core is drilled. This project is already under way at a site called North-GRIP, about one hundred fifty kilometers farther north along the centerline of the ice sheet, where scientists hope the Eemian layers will tell a clearer tale.

As Claire Parkinson has so often observed, it is the nature of scientific inquiry that as more and more data are assembled, some of yesterday's near-certainties prove to be in error while some of today's most tentative speculations prove true. So it is and so it is sure to be with the study of the Earth's climate.

Wally Broecker, in commenting on the current flurry of speculation regarding atmospheric oscillations and their effect on the operation of the Great Conveyor Belt, also advises caution. Although he finds such speculation "interesting," he feels it is premature to think that scientists have identified a mechanism that may explain how the Conveyor system relates to the sudden, dramatic changes that have been observed in the ice core record. "Remember," he warns, "our ignorance about deep water formation is extraordinary," adding that we are going to need huge quantities of additional data and many more years of study before we find all the answers we are looking for.

The message from every one of these scientists is the same: our knowledge about the mechanisms that drive the Earth's climate is large—but our ignorance about them is far larger.

Together, our species is conducting a massive, unintended

experiment on this planet. The evidence is mounting that we are altering basic chemical balances—perhaps irreversibly—even as we don't yet understand what these alterations may mean. We are in a race against ourselves, struggling to understand a highly complex system. As signs of changing climate patterns around the Earth continue to grow, the search for a definitive climate signal that will catalyze our collective response becomes ever more urgent. Perhaps such a signal will someday emerge from the Arctic ice. Perhaps it already has.

# Notes and Sources

Introduction

p. 7 *There had been talk:* Jim Hansen's speculations during the late 1980s about the advent of global climate warming are summarized in Andrew Revkin, *Global Warming: Understanding the Forecast* (New York: Abbeville Press, 1992), 60. For Hansen's own discussion of the Goddard model results, see J. Hansen et al., "Global Climate Changes as Forecast by Goddard Institute for Space Studies (GISS) 3-Dimensional Model," *Journal of Geophysical Research* 93, no. D8 (August 20, 1988): 9341–64.

p. 10 *There is no such debate:* Michael McElroy comments on the reality of the greenhouse effect in "Trouble in the Greenhouse," *Harvard,* March–April 1994, 27.

p. 11 *When Washington was inaugurated:* For statistics on

increases in atmospheric $CO_2$ since the beginning of the industrial revolution, see ibid., 27.

p. 12 *The reason, quite simply:* For a brief general discussion of the role of sea ice as a climatic accelerator, see Per Gloersen et al., "Sea Ice Processes and Climate," in *Arctic and Antarctic Sea Ice, 1978–1987,* NASA publication SP-511 (1992), 1ff. Also see Tamira Shapiro Ledley, "The Impact of Snow and Sea Ice Variations on Global Climate Change," in *Proceedings: International Conference on the Role of the Polar Regions in Global Change,* vol. 1 (University of Alaska, 1992), 321–24.

p. 14 *The first surprise:* The LDB cold spot anomaly is explored in depth in chapter 3. For remote-sensed measurements of the sea ice anomaly since 1978, see Claire L. Parkinson and D. J. Cavalieri, "Arctic Sea Ice, 1973–1987: Seasonal, Regional, and Interannual Variability," *Journal of Geophysical Research* 94, no. C-10 (October 15, 1989): 14499–523. For an identification of the anomaly since the early 1960s, see J. R. Marko et al., "Implications of Global Warming for Canadian East Coast Sea Ice and Iceberg Regimes over the Next 50 to 100 Years," Canadian Climate Centre, report 91-9 (Downsville, Ont.: Atmospheric Environment Service, 1991).

p. 14 *According to earlier data:* See J. R. Marko and D. B. Fissel et al., "Iceberg Severity off Eastern North America: Its Relationship to Sea Ice Variability and Climate Change," in press, 21ff.

p. 14 *"the problem of the east Canadian/West Greenland cold spot":* Jim Hansen, personal communication, April 1993.

## I. NORTHEAST

p. 34 *As its name suggests:* An in-depth discussion of sea ice types can be found in Claire L. Parkinson et al., *Arctic Sea Ice,*

*1973–1976: Satellite Passive Microwave Observations,* NASA publication SP-489 (1987), 2–16.

p. 35 *The only kind of ice:* Ross D. Brown and Phil Cote, "Interannual Variability of Landfast Ice Thickness in the Canadian High Arctic, 1950–1989," *Arctic* 45, no. 3 (September 1992), 273–84. For a contrasting view, see Peter Wadhams, "Variations in Sea Ice Thickness in Polar Regions," in *Proceedings: International Conference on the Role of Polar Regions in Global Change,* vol. 1 (University of Alaska, 1990), 5.

p. 36 *The last type of ice depicted:* The description of icebergs in this and the following paragraphs is derived in part from *Icebergs,* a publication of the Canadian Atmospheric Environment Service Ice Centre, Ottawa, Ontario.

p. 43 *A look at a modern current chart:* A simplified and highly readable discussion of ocean currents in the LDB area is contained in the Canadian Atmospheric Environment Service publication *Baffin Bay and Davis Strait,* AES Ice Centre, Ottawa, Ontario.

p. 44 *Surface currents in the LDB area.* Current data from U.S. Pilot Chart for the northern North Atlantic, June edition (Washington, D.C.: Defense Mapping Agency).

p. 45 *Davis obviously had:* For a description of the Arctic voyages of John Davis, see Samuel Eliot Morison, *The European Discovery of America: The Northern Voyages* (New York: Oxford University Press, 1971), 583–616. Also see Paul-Emile Victor, *Man and the Conquest of the Poles,* trans. Scott Sullivan (New York: Simon and Schuster, 1963), 70–72.

## II. NORTH NORTHEAST

p. 61 *Climatologists have long understood:* For a general discussion of the history of polar sea ice research and the importance of the new satellite imaging technologies, see R. G. Barry et al., "Advances in Sea Ice Research Based on Remotely Sensed Passive Microwave Data," *Oceanography* 6, no. 1 (1993): 4–12.

p. 62 *Over the past several years:* Claire L. Parkinson, "Sea Ice as a Potential Early Indicator of Climate Change," in *First North American Conference on Preparing for Climate Change* (1987), 118–23. Also see C. L. Parkinson, "Spatial Patterns of Increases and Decreases in the Length of the Sea Ice Season in the North Polar Region 1979–1986," *Journal of Geophysical Research* 97, no. C9 (September 15, 1992): 14377–388.

p. 64 *The satellite record for this region:* Claire L. Parkinson and A. Gratz, "On the Seasonal Sea Ice Cover of the Sea of Okhotsk," *Journal of Geophysical Research* 88, no. C5 (March 30, 1983): 2793–802.

p. 65 *By 1990 the verdict was in:* Claire L. Parkinson, "The Impact of the Siberian High and Aleutian Low on the Sea Ice Cover of the Sea of Okhotsk," *Annals of Glaciology* 14 (1990): 226–29.

p. 68 *As quick as Parkinson had been:* Claire L. Parkinson, "On the Value of Long-Term Satellite Passive Microwave Data Sets for Sea Ice/Climate Studies," *Geo Journal* 18, no. 1 (1989): 9–20.

p. 68 *In answer to my questions:* C. L. Parkinson and D. J. Cavalieri, "Arctic Sea Ice, 1973–1987: Seasonal, Regional, and Interannual Variability," *Journal of Geophysical Research* 94, no. C10 (October 15, 1989): 14499–523.

p. 69 *A question that I asked:* For "increase in global atmospheric temperatures" and "decreasing stratospheric temperatures," see Gloersen et al., "Sea Ice Processes and Climate," NASA publication SP-511, viii; for "warming of the Alaskan permafrost," see P. Gloersen and William J. Campbell, "Recent Variations in Arctic and Antarctic Sea Ice Covers," *Nature* 352 (July 4, 1991): 33; for "sea level rise," see Malcolm W. Brown, "Satellite Data Indicates Sea Levels Are

Rising around the World," *New York Times,* December 20, 1994.

III. STRAIT OF BELLE ISLE

p. 89 *Ewing's dilemma involved:* William Wertenbaker, *The Floor of the Sea* (Boston: Little, Brown, 1974), esp. 103ff.

p. 96 *In their 1991 study:* Marko et al., "Implications of Global Warming."

p. 97 *A simple computation:* For this and the following paragraph, see ibid., 34–35.

p. 98 *Beginning in 1973:* AES Ice Centre forecaster Phil Cote, personal communication, April 1993.

p. 100 *Marko himself makes a similar point:* J. R. Marko et al., "Study of Interannual Variability of Sea Ice off the Canadian East Coast," vol. 1, unpublished study for Atmospheric Environment Service, Downsville, Ontario, May 1994.

IV. NORTH

p. 120 *First proposed in a series:* Wallace Broecker et al., "Does the ocean-atmosphere system have more than one stable mode of operation?" *Nature* 313 (May 2, 1985): 21–26; also see Wallace S. Broecker, "The Biggest Chill," *Natural History,* 91 (November 1987): 74–82.

p. 122 *As luck would have it:* The following description of Wally Broecker and his ideas derives from three sources: two lengthy telephone interviews with Broecker himself conducted in March 1994 and April 1995 and a published interview by Lawrence Lippsett, "Wallace Broecker '53: The Grand Master of Global Thinking," *Columbia College Today,* Spring/Summer 1992, 19–25. Reprinted by permission of Lawrence Lippsett and *Columbia College Today.*

p. 122 *But Wally Broecker is more than:* Comments by Richard Fairbanks are from Lippsett, "Wallace Broecker," 19.

p. 123 *Broecker has been recognized:* Important honors awarded Broecker since 1994 include the National Medal of Science (1996) and the Blue Planet Prize from the Asahi Glass Foundation, Tokyo (1996).

p. 123 *"I like to work":* Lippsett, "Wallace Broecker," 23.

p. 123 *In the case of the Great Conveyor Belt:* Wallace Broecker, personal communication, April 1995.

p. 125 *As the diagram suggests:* Broecker's latest and most comprehensive explanation of the Conveyor Belt hypothesis can be found in Broecker, "Chaotic Climate," *Scientific American,* November 1995, 62–68. This article was provided to the author in draft form and serves as the basis for much of the description that follows.

p. 125 *"a sluggish mass":* This phrase is from W. S. Broecker, "The Great Ocean Conveyor," *Oceanography* 4, no. 2 (1991): 79.

pp. 126–27 The diagram first appeared in Wallace S. Broecker, "The Biggest Chill," *Natural History,* 91 (November 1987): 74. Published by permission of the artist, Joe LeMonnier.

p. 128 *"I'm going to use a unit":* Lippsett, "Wallace Broecker," 20.

p. 129 *"In the North Pacific":* Ibid., 21.

p. 130 *"a system with multiple identities":* Ibid. For a discussion of the "salt oscillator," see Broecker, "Great Ocean Conveyor," 87. Also see W. S. Broecker et al., "A Salt Oscillator in the Glacial Atlantic? 1. The Concept," *Paleoceanography* 5 (1990): 469–77.

p. 134 *"Scientists don't yet have":* Wallace Broecker, personal communication, April 1995.

p. 135 *"There are those that believe":* Lippsett, "Wallace Broecker," 24.

p. 135 *When Broecker is asked:* For this and the following paragraph, see Broecker, "Great Ocean Conveyor," 88.

p. 135 *The bottom line:* Lippsett, "Wallace Broecker," 21, 24.

## V. KANGIA: THE ICE RIVER

p. 154 *Perhaps the easiest way:* The substance of Richard Alley's thesis on the Arctic as "climate control center" was provided to the author in draft form. It has since been published as "Resolved: The Arctic Controls Global Climate Change," *Arctic Oceanography* 49 (1995): 263–83.

p. 155 *"In a community":* Ibid., 279.

p. 156 *Richard Alley's scientific involvement:* Richard Alley, personal communication, December 1994.

p. 157 *Once he'd completed his Ph.D. thesis:* R. B. Alley and D. R. MacAyeal, "West Antarctic Ice Sheet Collapse: Chimera or Clear Danger?" *Antarctic Journal of the U.S.* 28 (1993): 503–11.

p. 158 *The ice core records are complex:* A discussion of the ice age climate dynamics as summarized in this paragraph can be found in W. S. Broecker and G. H. Denton, "What Drives Glacial Cycles?" *Scientific American,* January 1990, 49–56.

p. 159 *"No coupled climate model":* Alley, "Resolved," 279.

p. 160 *The story of these Bond-Heinrich cycles:* See H. Heinrich, "Origin and consequences of cyclic ice rafting in the northeast Atlantic Ocean during the past 130,000 years," *Quaternary Research* 29, (1988): 143–52. Also see Gerard Bond, Hartmut Heinrich, Wallace Broecker, et al., "Evidence for Massive Discharges of Icebergs into the North Atlantic Ocean during the Last Glacial Period," *Nature* 360 (November 19, 1992): 245–49.

p. 161 *Other researchers began:* Wallace Broecker, Gerard Bond, et al., "Origin of the Northern Atlantic's Heinrich Events," *Climate Dynamics,* Spring 1992, 265–73.

p. 162 *It was Broecker's colleague:* Gerard Bond, Wallace

Broecker, et al., "Correlations between climate records from North Atlantic sediments and Greenland ice," *Nature* 365 (September 9, 1993): 143–47.

p. 163 *An answer to this final part:* R. B. Alley and D. R. MacAyeal, "Ice-rafted debris associated with binge/purge oscillations of the Laurentide ice sheet," *Paleoceanography* 9 (1994): 503–11.

p. 164 *The answer, Alley found:* R. B. Alley, "Resolved," 271ff.

p. 169 *In light of such findings:* For the "new source of evidence" described in this and the following paragraphs, see M. Anklin et al., "Climate Instability during the Last Inter-glacial Period Recorded in the GRIP Ice Core," *Nature* 364 (July 15, 1993): 203–7.

p. 170 *The potential importance:* See especially J. W. C. White, "Don't Touch That Dial," *Nature* 364 (July 15, 1993): 186.

## VI. SOUTH

p. 188 *"one of the most persistent":* Robert R. Dickson et al., "The 'Great Salinity Anomaly' in the Northern North Atlantic, 1968–1982," *Progress in Oceanography* 20 (1988): 103.

p. 190 *"The most recent model":* See Sirpa Hakkinen, "Simulated Interannual Variability of the Greenland Sea Deep Water and Its Connection to Surface Forcing," *Journal of Geophysical Research* 100, no. C3 (March 15, 1995): 4761–70. Hakkinen shared a draft of this work with the author in 1994.

p. 190 *The scientific discovery:* The following pages contain a summary of Dickson et al., " 'Great Salinity Anomaly,' " 103–51.

p. 190 *Dooley and several colleagues:* H. D. Dooley et al., "Abnormal hydrographic conditions in the north-east Atlantic during the nineteen-seventies," *Rapport et Procès verpaux des Réunions du Conseil Permanent International pour L'Exploration de la Mer* 34 (Copenhagen, 1977), 56–59.

p. 190 *The next scientific observers:* R. T. Pollard and S. Pu, "Structure and circulation of the upper Atlantic Ocean northeast of the Azores," *Progress in Oceanography* 14 (1985): 443–62.

p. 193 Map: This illustration of surface currents for the subpolar gyre is based on a somewhat more complex "Transport scheme for the 0–1000 meter layer of the northern North Atlantic," first published in G. Dietrich et al., *General Oceanography,* 2nd ed. (New York: Wiley, 1975). The dates and locations for the salinity minimum are from Dickson et al., " 'Great Salinity Anomaly,' " 112.

p. 197 *"If you're asking about what might be called":* K. Aagard and E. C. Carmack, "The role of sea ice and other fresh water in the Arctic circulation," *Journal of Geophysical Research* 94, no. C10 (October 15, 1989): 14485–98.

p. 198 *Here, Hakkinen suggested:* Sirpa Hakkinen, "An Arctic Source for the Great Salinity Anomaly: A Simulation of the Arctic Ice-Ocean System for 1955–1975," *Journal of Geophysical Research* 98, no. C9 (September 15, 1993): 16397–410. Also see Hakkinen, "Simulated Interannual Variability."

p. 199 *"The GSA was probably the jolt":* For additional evidence, see Peter Schlosser et al., "Reduction of Deepwater Formation in the Greenland Sea during the 1980s: Evidence from Tracer Data," *Science* 251 (March 1, 1991): 1054–56.

## VII. Southwest

p. 219 *Our traditional mathematics:* Two excellent discussions of the scientific study of complex systems are found in

James Gleick's *Chaos* and Mitchell Waldrop's *Complexity* (see Bibliography).

p. 220 *There is another sense:* Waldrop, *Complexity*, 333.

EPILOGUE

p. 244 *Perhaps the most important event:* J. T. Houghton et al., eds., *Climate Change 1995: The Science of Climate Change, Contribution of Working Group I to the Second Assessment Report of the Intergovernmental Panel on Climate Change* (New York: Cambridge University Press, 1996).

p. 244 *The data from NASA's most recent generation:* The NASA program for ice imaging since mid-1987 has employed a pair of Special Sensor Microwave Imagers (SSM/Is) mounted aboard Defense Meteorological Satellite Program (DMSP) satellites. The first SSM/I provided data from 1987 through 1991. The second has provided data from 1991 to the present.

p. 245 *A team of Norwegian scientists:* Ola M. Johannessen et al., "The Arctic's Shrinking Sea Ice," *Nature* 376 (July 13, 1995): 126. Also see "Polar Sea Ice on the Wane," *Science News* 148 (August 19, 1995): 123.

p. 245 *On the other hand, preliminary analysis:* One difference between the analysis of the Johannessen group and that of NASA scientists derives from issues of intercalibration between the data sets. In their published analysis, the Norwegians chose to treat the sets separately, whereas Parkinson and her colleagues at NASA preferred at that time to treat them continuously. Claire Parkinson, personal communication, February 1997.

p. 245 *The bottom line in either case:* LBD sea ice extents from telephone facsimile ice charts and analyses as issued weekly by the Navy/NOAA Joint Ice Center, Suitland, Maryland, and from daily ice reports from AES Ice Centre Environment Canada. Reports on fast ice bridging in Smith

Sound from personal communications between April 1992 and March 1997 with Phil Cote, Claude Dicaire, and Don Coleman, all forecasters at AES Ice Centre Environment Canada.

p. 245 *Recent evidence indicates:* Sirpa Hakkinen, personal communication, March 1997.

p. 245 *A new generation of free-floating instrument platforms:* Raymond W. Schmitt, "ALACE, PALACE, Slocum: A Dynasty of Free Floating Oceanographic Instruments," *Oceanus* 39, no. 2 (1996): 6–7.

p. 245 *long-term atmospheric oscillations:* These refer to a pattern of decadal (twenty-to-thirty year) shifts in mean atmospheric pressures over the study area called the North Atlantic Oscillation (NAO). Current speculation links the NAO to patterns of deep convection in the northern Atlantic and to events such as the Great Salinity Anomaly.

p. 245 *The study of ancient climates:* Richard Alley et al., "Comparison of deep ice cores," *Nature* 373 (February 2, 1995): 393–94. Information about North-GRIP is from Richard Alley, personal communication, March 1997.

p. 246 *Wally Broecker, in commenting:* Wallace Broecker, personal communication, March 1997.

# Bibliography

The following list of titles is offered as an aid to the reader and is not in any way intended to be exhaustive. Among the hundreds of technical articles and papers that have been part of the research for this book, the most important are cited in the "Notes and Sources" section.

The titles listed here are intended as suggestions for the general reader who may be interested in pursuing particular topics in more detail. All have been important to the genesis and development of ideas in this book.

Cohen, Joel E. *How Many People Can the Earth Support?* New York: Norton, 1995. The definitive work on the global population problem written by one of the world's foremost theoretical biologists and containing a discussion of one of the key topics in any investigation of global climate change.

Firor, John. *The Changing Atmosphere: A Global Challenge.* New Haven: Yale University Press, 1990. A balanced scientific discussion of atmospheric chemistry and its relationship to changing Earth climate—a book that avoids histrionics and political hyperbole and is all the more powerful because of this.

Gleick, James. *Chaos: Making a New Science.* New York: Viking, 1987. The classic popular description of the emerging sciences of complexity.

Gordon, Anita, and David Suzuki. *It's a Matter of Survival.* Cambridge: Harvard University Press, 1991. A well-researched review of some of the Earth's most pressing environmental emergencies.

Gribbin, John. *Hothouse Earth: The Greenhouse Effect and Gaia.* New York: Grove Weidenfeld, 1990. A presentation of the evidence in favor of greenhouse-induced climate change, in combination with a discussion of the Gaia or "living Earth" hypothesis of atmospheric chemist James Lovelock and others.

Gurney, R. J., J. L. Foster, and C. L. Parkinson, eds., *Atlas of Satellite Observations Related to Global Change.* New York: Cambridge University Press, 1993. A broad-ranging collection of scientific papers about global change accompanied by some of NASA's most dramatic photographs of the Earth from outer space.

Kent, Rockwell. *N by E: A Record of a Voyage.* New York: Brewer and Warren, 1930: Middletown, Conn.: Wesleyan University Press, 1978. A collection of lyrical reflections about a sailing voyage to western Greenland in 1929, illustrated with a series of original woodcuts by the author and recognized as one of the classics of its genre.

Lopez, Barry. *Arctic Dreams: Imagination and Desire in a North-*

*ern Landscape.* New York: Scribner's, 1986. An evocation of Arctic landscapes and human aspirations, imaginings, and perceptions of them.

MacMillan, Donald B. *Etah and Beyond, or Life within Twelve Degrees of the Pole.* Boston: Houghton Mifflin, 1927. Captain Mac's book-length chronicle of his second expedition aboard schooner *Bowdoin,* 1923–24, including the tale of a winter spent frozen in the ice in northern Greenland.

MacMillan, Miriam. *Green Seas and White Ice.* New York: Dodd, Mead, 1948. An excellent account of some of *Bowdoin*'s exploits as written by Captain Mac's wife, Miriam. This is Arctic adventuring from a refreshingly feminine point of view.

McKibben, Bill. *The End of Nature.* New York: Random House, 1989. An impassioned wake-up call to members of our species about what we are doing to change the chemistry of the planet we live on.

Morison, Samuel Eliot. *The European Discovery of America: The Northern Voyages.* New York: Oxford University Press, 1971. A history of the voyages of John Davis (and others) as told by the most highly regarded American maritime historian of our time—a man who was also an accomplished small-boat sailor in his own right.

Mowat, Farley. *Sea of Slaughter.* Boston: Atlantic Monthly Press, 1984. A cry of anguish for the disappearing species, both aquatic and terrestrial, of the Saint Lawrence Gulf and the seacoasts of eastern Canada. In many ways this book is the masterwork of this celebrated Canadian naturalist.

Parkinson, Claire L. *Earth from Above: Using Color-Coded Satellite Imagery to Examine the Global Environment.* Mill Valley, Calif.: University Science Books, 1997. A highly readable and lavishly illustrated introduction to the use of color-coded satellite imagery in examining features of the

Earth's environment, including an excellent chapter on sea ice.

Revkin, Andrew. *Global Warming: Understanding the Forecast.* New York: Abbeville Press, 1992. Sponsored by the American Natural History Museum and the Environmental Defense Fund, this is a summary of the standard arguments, pro and con, for greenhouse-induced climate change, along with an excellent collection of photographs and images on the subject.

Waldrop, M. Mitchell. *Complexity: The Emerging Science at the Edge of Order and Chaos.* New York: Simon and Schuster, 1992. The book that carries on from the place that Gleick's *Chaos* ends. A readable story about Nobel Prize–winning physicist Murray Gell-Mann and others and the founding of the Santa Fe Institute.

Wertenbaker, William. *The Floor of the Sea: Maurice Ewing and the Search to Understand the Earth.* Boston: Little, Brown, 1974. A good general treatment of Maurice Ewing and the story of the discovery and verification of the theory of plate tectonics.

# Acknowledgments

When a book is also a voyage, many people help to ensure that its course is true. Some of these have also been part of the book and are already familiar to the reader. Others, unnamed or mentioned only in passing, are equally important for contributing their ideas, sharing their written work, critiquing portions of the manuscript, shepherding various phases of the project to completion.

The book would never have happened without the help of two people: Claire Parkinson and Carol Young. During the nearly five years of research and writing, Claire served in various roles as teacher, critic, editor, and encouraging voice, just as each was needed. To the extent that the scientific content of this book is fair and evenly presented, she is due much of the credit. To the extent that it is not, the author and only he shall shoulder the blame.

Carol Young was also there from the beginning. She edited text and created graphics for the fundraising campaign, and she produced eleven of the maps that illustrate this text. Even more important, she worked an editorial miracle when the book's manuscript was still unkempt, cutting, organizing, shaping, until the book came to life and the story emerged from the written page as dramatically as it had happened.

During the earliest stages of the project, Hubert Jessup was a valuable source of advice and guidance. So were Ted Brainard, Rennie Stackpole, Bob Rice, Chris Knight, Susan Kaplan, and Bernadette Bernon, all of whom also served as members of the project's Advisory Panel. Frank Kniskern, director of the Navy/NOAA Joint Ice Center, and Dave Mudrey, chief of AES Ice Centre Environment Canada, both provided valuable technical assistance, as did Canadian ice forecasters Phil Cote and Bob Tessier.

Major Willy Kerr, Newbold Smith, and Captain Andy Chase all shared their knowledge of the sailing routes and remote anchorages of the west Greenland coast. Girt Saaby, Erik Moeller, and Henry Fuller provided useful assistance en route. John Griffiths and my wife, Kay, both sailed as crew members during four weeks of the voyage, unmentioned in the book for reasons of narrative economy, and John served as a much-needed peacemaker among the crew.

Many individuals and organizations supported the project with donations of money or gifts-in-kind to the sponsoring organization, Associated Scientists at Woods Hole. A special thank you goes to the William Lyon Phelps Foundation for its generous support, as well as to Mustang (USA) Corp., Ted and Liz Brainard, Peggy Kling, John Hulseman, Gary and Melissa Brown, and the author's parents, Charles and Betsy Arms.

Among the scientists who appear in this book, John Marko gave unselfishly of his time and ideas, as did Richard Alley, Sirpa Hakkinen, and Wally Broecker. Dick Campbell and Red Wright provided support from Woods Hole and aided importantly in the fundraising effort. Bill Fitzhugh and Michael McElroy were both more influential in the evolution of ideas herein than they may have known.

I owe a special debt of gratitude to Mike Auth, whose eagerness to sail to the ice was one of the important motivating forces in the project. I'm indebted as well to Amanda, Pete, and Blue for putting up with my goal-directedness and my all too frequent insensitivites to them and for doing whatever was required to take *Brendan's Isle* to the ice and safely home again.

I wish to thank my editor at Anchor Books, Rob McQuilkin, for his invaluable help in the final shaping of the manuscript as well as my agent, Arnold Goodman, for his unfailing enthusiasm for the project. Jan Scott and John Miller also added importantly to the editorial process with early critical readings of the manuscript. But the most important editor—the one who was there from first word to last—was my wife, Kay. Page by page she encouraged, prodded, criticized, cajoled. She abhorred the extra word and forced me to say exactly what I meant. She was and is the ultimate reader.